SPATIAL DATABASE

TRANSFER STANDARDS:

Current International Status

SPATIAL DATABASE TRANSFER STANDARDS:

Current International Status

Edited by

H.MOELLERING

ICA Working Group on Digital Cartographic Transfer Standards

Published on behalf of the
INTERNATIONAL CARTOGRAPHIC ASSOCIATION

by

ELSEVIER APPLIED SCIENCE
LONDON and NEW YORK

ELSEVIER SCIENCE PUBLISHERS LTD
Crown House, Linton Road, Barking, Essex IG11 8JU, England

Sole Distributor in the USA and Canada
ELSEVIER SCIENCE PUBLISHING CO., INC.
655 Avenue of the Americas, New York, NY 10010, USA

WITH 8 TABLES AND 18 ILLUSTRATIONS

British Library Cataloguing in Publication Data

Spatial database transfer standards:
current international status.
I. Moellering, H.
III. International Cartographic Association
910.285574

ISBN 1-85166-677-X

Library of Congress CIP data applied for

Printed in Great Britain at the University Press, Cambridge.

PRESIDENTIAL FOREWORD

The impact of the computer on cartography is now all pervasive. As a result, many nations hold much of their cartographic data in digital format. This has made the issue of spatial database transfer standards one of great significance for the discipline of cartography. This volume is the most definitive study on the topic yet produced, and is an excellent example of the international scientific cooperation which marks the work of the International Cartographic Association (ICA).

The ICA was established in 1959 in Bern, Switzerland, and currently has 62 member countries. The Association is the world body for cartography and much of its scientific work is carried out through its various Commissions and Working Groups. This volume is a result of the efforts of a working group of the Standing Commission on Advanced Technology. The group was chaired by Professor Harold Moellering who, with the assistance of Hedy Rossmeissl of the United States Geological Survey, edited this volume on *Spatial Database Transfer Standards*. The current situation in 16 ICA member nations is discussed together with the efforts of two international groups working on standards issues. The production of this volume is a major contribution to the fulfillment of one of ICA's central aims which is the initiation and coordination of cartographic research through cooperation among cartographers from all member nations and, as President of ICA, I welcome its appearance.

D.R.F. Taylor
President
International Cartographic Association

PREFACE

The growth in concern over environmental issues, at scales ranging from the local to the global, has been matched by the rapid evolution of geographical information systems in terms of power and functionality. The systems provide a tremendous capability to manipulate and analyse geographical data. As the application of these systems is taken up by an ever growing community of users, there is one common need—data! Industry estimates now suggest that 60 percent of the cost of implementing a geographical information system (GIS) involves the development of the necessary data base.

While each GIS application may require certain unique data layers, many themes will be common to a number of applications. Many applications in local government environments depend upon multiple agencies for data base development. For these efforts to be successful, data standards are essential. Such standards must address a range of issues more encompassing than exchange formats. Digital data standards are especially key to success in the various global change research projects now underway.

Thus, it is particularly timely that Dr Moellering has been able to assemble an international summary of standards work. He is to be congratulated for a thorough and wide-ranging volume addressing standards issues from an international perspective. As Chairman of the ICA Standing Commission on Advanced Technology, I offer thanks to Dr Moellering and his colleagues for a valuable contribution.

K. Eric Anderson
Chairman
ICA Standing Commission on Advanced Technology

ACKNOWLEDGEMENTS

The ICA Working Group on Digital Cartographic Database Exchange Standards was formally founded at the ICA meeting held at Budapest in August, 1989. At that meeting a goal was established for the Working Group to write a monograph concerning activities in the cartographic world to develop spatial database transfer standards. This monograph is the result. It contains 18 chapters that describe these standards activities on a world wide basis.

The editor offers his thanks to each of the authors for their efforts devoted to developing their chapters that describe the activities in their countries or organizations. As part of this writing effort the Working Group held a meeting in Switzerland in July, 1990, to review the draft chapters. The final copies of the chapters were finished in the Fall of 1990.

Thanks are due to the Executive Secretary of the Working Group, Ms Hedy Rossmeissl of the US Geological Survey in Reston, Virginia, who supervised the assembling and formatting of the chapters into the monograph. She also supervised the production of the final camera ready copy. Secretarial support for this project was provided by Adonnis Goldstein and Cathy Taylor. Without their dedication and hard work this document would not be a reality.

HAROLD MOELLERING,
Editor
Columbus, Ohio, USA

CONTENTS

APPROACHES TO SPATIAL DATABASE TRANSFER STANDARDS: AN INTRODUCTION

Prof. Harold Moellering
Dept. of Geography 103 BK
Ohio State University
Columbus, Ohio
U.S.A. 43210
Bitnet: TS0215@OHSTVMA

INTRODUCTION

Since the middle of the 1960's various individuals, groups and organizations in many countries in the world have been building various kinds of cartographic databases to use with their cartographic software systems to analyze and display their data. In the early years this work was uncoordinated and somewhat haphazard, but as the years progressed it came to be realized that great efficiencies could be gained if the cartographic database built by one group could be used by another on a system that is different from the originating system. It has been further realized that more efficiencies could be gained if transfer standards could be developed that would facilitate transfers of cartographic databases.

This idea has arisen in many cartographic and spatial data processing groups and organizations in the 1980's where groups have been working on this problem for some years. Other groups and organizations have become interested in the challenge of this problem more recently. As this work was initiated in various countries, research workers began to informally compare notes and experiences concerning such developments. A number of informal discussions have been held at the last few International Cartographic Association technical meetings. During the ICA congress in Morelia, Mexico, in 1987 under the Commission on Advanced Technology under the leadership of Dr. Eric Anderson, the topic was formally examined and discussed.

From those meetings the ICA Working Group on Digital Cartographic Database Exchange Standards was founded in early 1989 by Prof. Harold Moellering of the U.S.A. The following goals were established by Prof. Moellering for the Working Group (WG):

1) The WG will be organized in the 1989 time period;

2) The initial meeting of the WG will be held at the ICA meetings in Budapest in August, 1989;

3) To exchange information and reports by the ICA member countries concerning the development of digital cartographic data exchange standards;

4) To collect and distribute in the WG copies of all standards published in ICA countries;

5) To serve as a focal point of information concerning digital cartographic data exchange developments throughout the world;

6) To identify research needs that arise from the standards process;

7) A presentation by each member of the WG will be made at the Budapest meetings concerning cartographic standards activities in his/her member country.

An effort was made to contact the ICA countries that were engaged in or interested in such work to nominate a member for the WG. International organizations that were known to be working on this problem were invited to nominate an observer to the WG. The idea was to have a member or observer to represent each active ICA nation and international organization in the world.

The founding meeting of the Working Group was held at the ICA technical meetings in Budapest, Hungary in August, 1989. Representatives from 16 ICA countries were present for this meeting. At this meeting many WG members gave presentations concerning the status of cartographic database transfer standards in their country. At this meeting in Budapest, the WG also added an eighth goal, that of producing an ICA monograph that discusses the present state of development of such standards in the various ICA countries and organizations throughout the world. That desire has resulted in this monograph coming into existence. Since the meetings in Budapest the members of the Working Group have been working diligently to produce their chapters. The WG met in Switzerland in July, 1990 to review their draft chapters and to do the final polishing on them. You see the results of their diligent efforts before you.

This monograph is organized into an introduction by the editor that provides the background for this effort. It also provides an explanation of some of the important concepts that underlie this effort, as well as a discussion of the transfer process itself. This preliminary material is important to the reader can more clearly understand the individual standards efforts. This introduction also contains a brief review of each following chapter to succinctly encapsules each effort in a larger context. It concludes with a brief summary of this introductory material. Following that are about 20 substantive chapters that presents the efforts going on in each country and organization represented. This presents a comprehensive picture of the cartographic database transfer standards development efforts that are active throughout the world. In one or two cases groups invited to write chapters did not complete them.

As one reads these individual descriptions of the activities of these standards developments in the ICA countries and other organizations, it is possible to get the impression that all of the theoretical details are completely understood. This is far from the case. Rather there are many theoretical concepts that are incompletely understood or in some cases, not well at all. These situations provide a number of interesting research opportunities, as described in Moellering (1991b).

BASIC CONCEPTS

In order to efficiently understand the transfer process and following discussions, it is crucial to clearly comprehend the fundamental cartographic theory that underlies this work. They are the notions of real and virtual maps, deep and surface cartographic structure, and the notion of cartographic data levels. These concepts, and many others, are emerging from the developing area of analytical cartography, which is major thrust to develop a more theoretical and mathematical basis for cartographic concepts. A brief review of analytical cartography has recently been written by Moellering (1991a).

The first major concept of interest here is that of real and virtual maps. During the early 1970's there arose many cartographic products such as CRT images and digital terrain models that went beyond the conventional definition of a map as a fixed hard copy product. These developments resulted in calls to expand the definition of what constituted a map. This dilemma was crystallized by Morrison (1974) in the lead article of the first issue of the American Cartographer where he recognized this growing problem and called for an expanded definition of what constitutes a map. Moellering was faced with the same problem and took up the challenge issued by Morrison. After a few years of research the concept of real and virtual maps was proposed (Moellering, 1980).

It turns out that there are two crucial characteristics that differentiate conventional hard copy maps from other kinds of virtual maps. The first is whether the product can be directly viewed as a cartographic image. Conventional maps and CRT images can be seen that way, but cartographic data files and things like Fourier transforms cannot. They must be transformed to a state that has direct viewability first. The second crucial characteristic is whether the product has a permanent tangible reality. Figure 1 shows a four class diagram that shows the resulting classes of real and virtual maps that are generated by yes/no answers to the two characteristics. The definitions for these classes (Moellering, 1980) are as follows:

Real Map - is any cartographic product which has a directly viewable cartographic image and has a permanent tangible reality (hard copy). There is no differentiation as to whether that real map was produced by mechanical, electronic means.

	DIRECTLY VIEWABLE AS A CARTOGRAPHIC IMAGE: YES	DIRECTLY VIEWABLE AS A CARTOGRAPHIC IMAGE: NO
PERMANENT TANGIBLE REALITY: YES	REAL MAP Conventional Sheet Map Globe Orthophoto Map Machine Drawn Map Computer Output Microfilm Block Diagram Plastic Relief Model	VIRTUAL MAP-TYPE 2 Traditional Field Data Gazeteer Anaglyph Film Animation Hologram(stored) Fourier Transform(stored) Laser Disk Data
PERMANENT TANGIBLE REALITY: NO	VIRTUAL MAP-TYPE 1 CRT Map Image a) refresh b) storage tube c) plasma panel Cognitive Map (two-dimensional image)	VIRTUAL MAP-TYPE 3 Digital Memory(data) Magnetic Disk or Tape(data) Video Animation Digital Terrain Model Cognitive Map (relational geographic information)

Figure 1. The Four Classes of Real and Virtual Maps

<u>Virtual Map - Type I</u> - has a directly viewable cartographic image but only a transient reality as has a CRT map image. This is what Riffe called a temporary map. Given the direction of current scientific work, electro-cognitive displays may be possible.

<u>Virtual Map - Type II</u> - has a permanent tangible reality, but cannot be directly viewed as a cartographic image. There are all hard copy media, but in all cases these products must be processed further to be made viewable. It is interesting to note that the film animation adds a temporal dimension to the cartographic information.

<u>Virtual Map - Type III</u> - has neither of the characteristics of the earlier classes, but can be converted into a real map as readily as the other two classes of virtual maps. Computer-based information in this form is usually very easily manipulated.

Here it can be seen that conventional cartographic products such as sheet maps, atlases, and globes that have a fixed tangible reality, and are directly viewable as a cartographic image are called Real Maps. The other three classes lack one or both of the two characteristics and are called Virtual Maps. These three classes provide for the expanded definition of maps that reflects many of the developments in modern cartography. It turns out that Virtual Maps can contain the same information as a Real Map, and in the case of a cartographic database, perhaps more. Moellering recognized that even cartographic databases should be considered maps because they can contain the information of a Real Map and can be transformed into one if necessary. This solution solves the dilemma recognized by Morrison.

Expanding on a notion pioneered by Tobler (1979), transformations between the four classes of Real and Virtual Maps can be used to define all of the important data processes in cartography. As shown in Moellering (1984) these 16 transformations define operations such as digitizing (R --> V3), CRT display (V3 --> V1), making a CRT hard copy (V1 --> R), direct plotting (V3 --> R), or data base transfer (V3 --> V3). These transformations can also be used to design cartographic systems (Moellering, 1983) and can also be used to define the field of cartography itself (Moellering, 1987).

It turns out that the information in these various classes of Real and Virtual Maps can be transformed from one to another as shown in Figure 2.

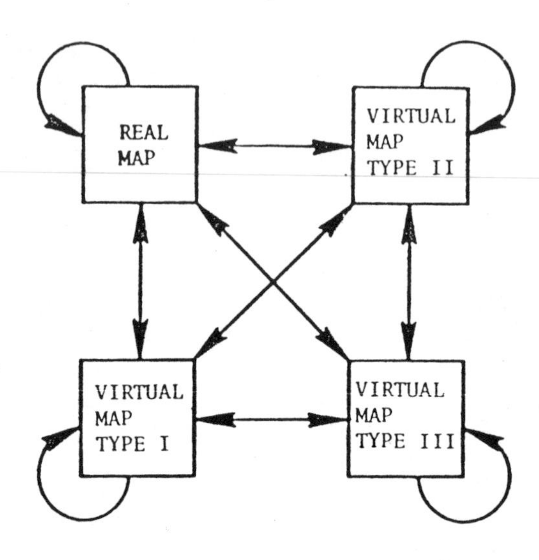

Real and Virtual Map Transformations

Figure 2. The 16 Real/Virtual Map Transformations

The second major concept is that of deep/surface cartographic structure developed by Nyerges in his dissertation (1980). Surface structure is defined as the graphic realization of cartographic information and when realized in a fixed hard copy form is called a Real Map. Following the precepts of Noam Chomsky in structural linguistics, Nyerges realized that there was a direct spatial analog to the linguistic deep structure. Cartographic deep structure is the set of spatial entities, attributes and relationships between them that may or may not be graphically realizable. This information is always in a virtual state and is usually found in Virtual Map Type 3 databases. Figure 3 shows this relationship graphically where the lower part of the figure, the surface structure made up of Real Maps and

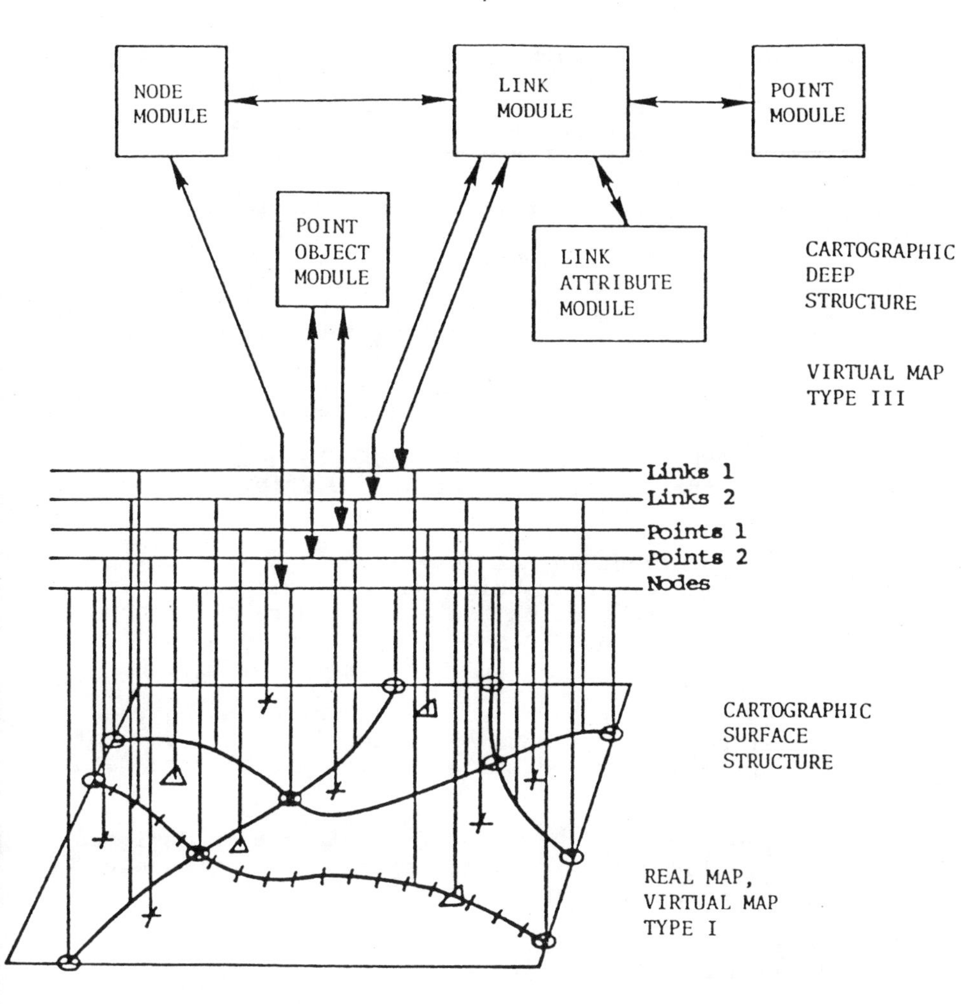

Figure 3. Deep and Surface Cartographic Structure

Virtual MapsType 1, is the area where traditional cartography has been focused for the many centuries of its long and glorious history. However, in the last decade, a new area has been identified, deep structure made up of Virtual Maps Type 3, that is now the focus of an expanding area of analytical work. Deep structure contains work on spatial data structures, analytical operators, fractals just to name a few. For a fuller discussion of analytical map use, see Nyerges (1991). This concept is introduced here because cartographic database transfers are really Virtual Map Type 3 deep structure transfers. More explanation and discussion will follow in a later section.

This concept greatly expands our understanding of

cartography and directly reinforces the concept of virtual maps. It can also be seen that analytical cartography largely operates in the area of deep structure and displays its results in a surface representation of some kind, many times in the form of Virtual Map Type 1 CRT images.

The third major concept of importance here is that of Nyerges data levels (1980) which define the levels of spatial data from the most general level of Data Reality to the most specific level of Machine Storage as the bits and bytes are contained in computer hardware systems. The following table lists these levels and provides a short description of them.

Table 1. Nyerges Data Levels

1) Data Reality - The real world and data pertaining to it concerning cartographic entities and relationships between them.

2) Information Structure - A formal model that specifies the organization of information pertaining to a specific phenomenon. It includes data classes and relationships between them and acts as a skeleton for the canonical structure.

3) Canonical Structure - A data model representing the inherent structure of a data set which is independent of specific applications and systems which manage such data.

4) Data Structure - A logical data organization designed for a particular system in which specific relationships and links are implemented.

5) Storage Structure - A specification of how a particular data structure is stored in data records in a particular system.

6) Machine Encoding - The physical representation of how the structure is held in the physical devices of computer system hardware.

It turns out that early developments in cartographic data structures many researchers tried to define the Data Structure, Nyerges Level 4, directly without having defined the more general levels of the Information Structure and the Canonical Structure first. The Information Structure is a formal model of the information to be housed and managed in the spatial database, while the Canonical Structure is a model of the data to be coded, structured and housed in the database. The Data Structure is dependent on those two more general levels if it is to be efficiently defined. The Data Structure is then stored in data records in the Storage Structure which is physically stored as bits and bytes in the computer hardware in Nyerges Level 6. This work suggests why so many spatial data structures of the 1970's did not work very efficiently or in some cases failed outright. Hence, this concept is crucial to understand in order to clearly see what is involved in the cartographic database transfer process.

THE DATABASE TRANSFER PROCESS

At the outset, one might conceive of this cartographic database transfer process as capturing and moving a Real Map image from one system to another. In this view such a transfer would be a surface structure transfer. However, the cartographic world is a far richer place than that. The cartographic information to be transferred must first be organized into a Virtual 3 deep structure database. This is because such a Virtual 3 deep structure organization allows for the addition of more deep structure information which is associated with the database, but is not of a surface structure (graphic) nature. Such deep structure information can include a wealth of attribute information, for example, that may be used in the database for analytical purposes and not necessarily for graphic rendering. The key idea here is that such a database transfer between systems is a deep structure Virtual 3 transfer and not a surface structure transfer. A deep structure transfer facilitates building the transfer file itself as well as the restructuring of the data itself into the spatial data structure of the system to which the database is being transferred. One must realize that when cartographic or spatial data is in a Virtual 3 deep structure form, it is much more flexible and manipulable than it would be in other states.

At this point it is important to understand the difference between the use of pairwise converters for database transfers, the current situation, and the use of a more general metafile transfer, the proposed situation for most national transfer standards. One issue involves the number of database converters that would be required to do the job. Suppose that there are 100 cartographic systems with heterogeneous data structures, software systems and hardware architectures in an open systems environment that are candidates for transferring spatial databases among them. Using the current situation of pairwise data converters as an example one would need to accomplish the transfer between

$$\frac{n\,(n-1)}{2} = \frac{100\,(99)}{2} = \frac{9{,}900}{2} = 4{,}950 \text{ pairwise converters}$$

all combinations of systems. The desired situation is to have a common metafile transfer standard that can be used for (almost) all database transfers where 2 n + = 2 (100) = 200+ metafile converters would be required to accomplish the task. It can be clearly seen that the metafile transfer approach is preferred because in theory about 200 metafile converters would be required, whereas 4,950 pairwise converters would be required. This is a tremendous difference. In practice this difference would probably be smaller because in the pairwise case not all combinations would require implementation, and in the metafile case converters would probably be required for general classes of spatial database systems. But even if three times more metafile converters were required, 600 in this example, and only 60 percent of the pairwise converters were required, the difference would be about 3,000 converters for the pairwise case and 600 for the metafile case. This is still a five to one savings for the metafile approach. That is why national standards groups are moving towards this solution for their database transfer requirements. For the long run, the metafile approach is clearly preferred.

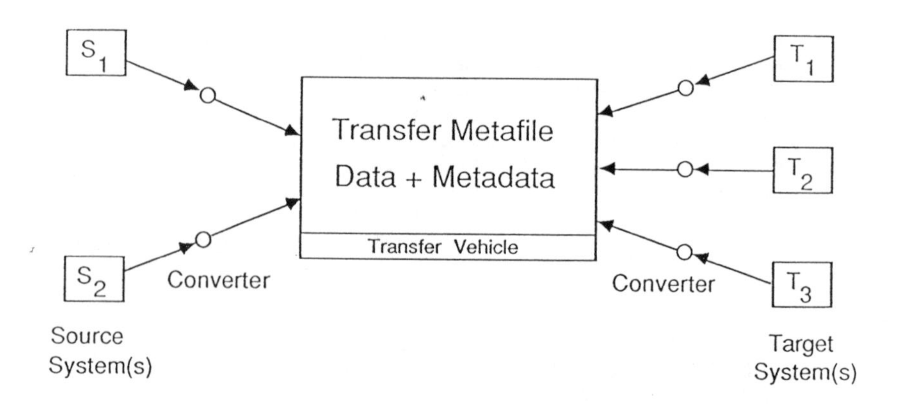

Figure 4. The Cartographic Database Transfer Process

Figure 4 shows an illustration of the general cartographic database transfer process. On the left is the Source Database with its own information structure, canonical structure, and data structure, data elements and modules. On the right are several Target Databases to which one desires to transfer the cartographic database. These cartographic databases are associated with different systems that may have very different data structures and hardware architectures. Therefore the source data structure usually cannot be transferred directly between the two systems. It is usually necessary to take the source database and convert it into the organization of the Transfer Metafile. This conversion is done by a software converter designed for that purpose that reads the data structure of the Source Database and converts it into the structure of the Transfer Metafile.

The Transfer Metafile has a structure that is specified by the national standard for which it was designed. This Transfer Metafile is the actual vehicle that carries the data from the Source to the Target systems. The internal structure of various national transfer standards differ widely in approach and design. They range from fairly straightforward formats to fairly rich transfer mechanisms. This will be discussed later. Many national Transfer Metafiles have as their base an international transfer standard such as ISO 8211 or others to act as base transfer vehicle. This vehicle facilitates the actual mechanics of the transfer.

When the Transfer Metafile arrives at the Target System, it must be converted from the organization of the transfer file to that of the target system(s). This is accomplished by a converter that reads the Transfer Metafile and converts the structure of the data to that of the Target System. That system itself will then insert the actual data into its database.

One question that arises from the above description is what kinds of data components can be transferred with such a process. Fundamentally, one has descriptions of features from the real and attributes associated with them, cartographic objects that contain the geometry and topology of the features, information on the data quality, and other essential information such as projection, coordinate system, etc. The features have been defined by the individual national groups and code such things as mountains, streams, transportation, cities, soil, population and a host of other things that exist in the real world. Associated with these features are a set of attributes that contain information about those features such as mountain elevations, stream flow, transportation type, city name, soil type, population components, and so on. The cartographic objects usually have a set of primitives that begin with the level of spatial dimension, 0-, 1-, 2-, 3-, etc., and may contain information about topological properties of the object, such as connectivity or contiguity (neighbor relation). Quality information that relates to the above features, attributes and objects is specified and carried along with the transfer. Depending on how this information is specified, this quality information could relate to large classes of data, or to individual features or objects themselves. Ancillary information such as projection, coordinate system is usually part of the transfer file and serves an important role in processing the data. Finally, some standards include metadata, which is data that relates to the substantive data being transferred, to aid in the transfer process. Such metadata can be particularly helpful to the decoding converter that receives the Transfer Metafile and converts into the structure of the Target System. The nature of this metadata varies greatly between the various national transfer standards. This above discussion also implies that each national standard has specified things like feature definitions, object definitions, quality levels, and the Transfer Metafile organization itself.

The Transfer Process and Data Models
The database transfer process can also be examined from the point of view of the data model(s) involved. At the conceptual level it involves the Information Structure, Nyerges level 2, and the Canonical Structure, Nyerges level 3, while at the operational level it involves the Data Structure, Nyerges level 4 and the Storage Structure, Nyerges level 4. The Information Structure is a model that describes how the entities from the real world that are contained in the database are related one to another. In order for a database transfer to take place effectively, the source database and the target database must be compatible with one another. For example, if the source database only contains information pertaining to roadway systems and the target database only information pertaining to streams and waterways, a data transfer is not feasible because the source and target databases are not compatible with each other. If, on the other hand, the target database contains data on streams, waterways, roadways, and communication linkages, then a transfer from the source to the target database is feasible. Therefore, at the information structure level, when the subject matter of the source and target databases is compatible, a transfer between them is usually feasible, assuming that the lower Nyerges levels are compatible too.

One can now look at the Canonical Structure, Nyerges level 3, which specifies the model of the data that is contained in a database. At the outset one must be clearly aware that there are three general families of data structures that are specified by such data models. They are vector, raster and relational. These are treated as distinct classes of data organization which are not compatible with one another. Therefore, it is not feasible to attempt to transfer data from a source database that has some sort of raster structure to a target database that has some sort of vector structure. Going between these three families of spatial data organization is a data conversion problem beyond the sort of spatial data transfer problem being discussed here. Since this discussion focuses on the data transfer question, such conversions between the three families of spatial data will not be further considered here. It is therefore assumed that database transfers of the kind being discussed here will be transfers <u>within</u> each of these three fundamental families of spatial data. However, with some of the national efforts, it is possible and feasible to transfer data from these three families within different sections of a transfer file. In that case raster data from the source system would arrive as raster data at the target system, vector from the source system would arrive at the target system as vector data, and so on. No conversions between families could be made. The intriguing question as to whether there may exist a unified theory of spatial data that would permit more general data models and structures has been raised and discussed in Moellering (1991b).

The data model specifies the formal organization of the spatial data for a particular system. When transferring data between two systems that have the identical data model, even when there are differences in the implementation of the Data Structure, Nyerges level 4, the transfer is straightforward. This current discussion is setting aside the question of hardware architecture differences for the moment, which will be discussed later. The more typical case is when the data models in the source and target systems are different, even though they are in the same data structure family. The general answer to this question is that if the source and target systems contain the equivalent information, even though it may be structured in differing data models, a data transfer can usually be carried out. However, there are limitations to this possibility. For example, if the source system data model contains geometry and topology for the fundamental cartographic objects, and the target system data model only contains geometry for such objects, the transfer can be made, although it will be an information losing transfer. However, if the source system data model only contains geometric information on the cartographic objects, and the target system data model contains both geometry and topology, then a problem arises because the target system requires information that is not found in the source system. Since the data transfer tools described below are not designed to generate new information for the target system, such a transfer cannot be carried out. Therefore, the general principle to be learned here is that these kinds of data transfers can be information losing, although it is inefficient to do so, but they cannot be information gaining.

One can then consider the specifics of the Data Structure, Nyerges level 4. It is very common for large differences to exist between Data Structures in the source and target systems even though the data models may be fairly similar. Obviously large differences will exist between the two Data Structures when the data models are also very different. Assuming that the information contained in the source and target systems are equivalent, the transfer process must restructure the data so it is compatible with the target system when they arrive at that end of the process. The transfer process must also overcome differences at the Storage Structure level, Nyerges level 5, as well. There will almost always be large differences between the Storage Structures of the source and target systems due to differences of the operating system and hardware architecture differences. These kinds of differences are not inconsequential and must be properly handled by the transfer process.

Transfer Machinery

One can now begin to turn to the machinery that actually carries out the data transfer. It turns out that the various national standard proposals differ considerably in approach and outlook to this problem. This section provides a conceptual framework of the range of the kinds of transfer machinery that is being proposed in the field. In all cases such transfer machinery must accomplish two major tasks:

1) It must transfer spatial data from source to target system where there are differences between the data models, data structures, and storage structures of the two systems, as discussed above;

2) It must transfer the spatial data in such a way that the process transcends differences in hardware architecture between the source and target systems.

This second requirement is just as important as the first. Every transfer standard must address this question. In general, approaches to the problem range from using a fairly fixed format to using one of the international standards such as ISO 8211 or EDIFACT as a basic implementation vehicle.

Reflecting back to Figure 4, every standard has a transfer file. Those that are more rigid transfer the information in the form of a fairly rigid data model. As the transfer standard becomes more flexible the data model reflected in the transfer file becomes more flexible. The most flexible form of transfer standard can be called a transfer mechanism. Here the transfer data model is a very flexible minimal model that can accommodate a wide range of source/target data models. This range of approaches to the problem can be conceptually viewed as a continuum as illustrated in Figure 5. At the extreme left one has the system specific fixed

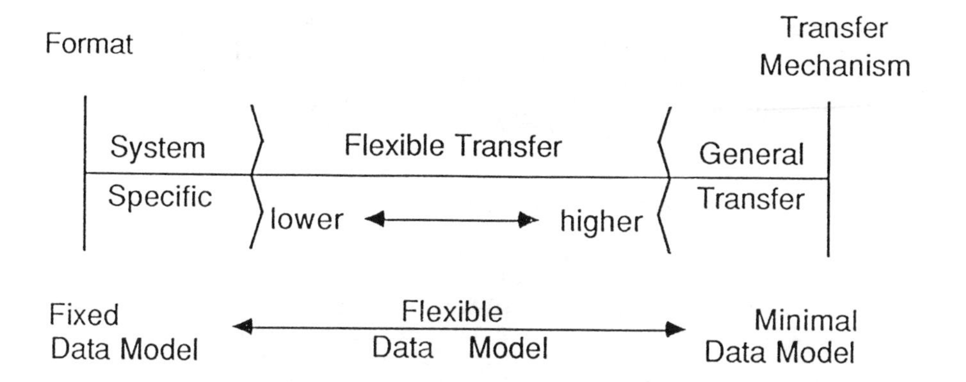

Figure 5. The Range of Cartographic Database Transfer Processes

data model that would be a fixed fairly rigid format. If the source and target systems were the same fixed data model used by the transfer standard, the data transfer would be very simple. However, this level is very inflexible can generally cannot accommodate a wide range of other data models for transfer. It would have fairly limited use between differing systems. Few if any of the national standards proposals are this rigid. More likely is the case where the transfer standard has a its own fairly specific data model that has been designed to accommodate a wider range to source and target data models. These more flexible data formats are more widely applicable to the situation where there are wider differences between the source and target data models. These systems require that the source data model be converted into the transfer data model which is then later converted into the target system data model. This also implies conversions in the data structures and file structures as well. The more flexible is one of these transfer formats, the wider range of source and target data models that can be accommodated. However, this implies increasing complexity of the standard as the flexibility of the format increases. The most flexible kind of a transfer standard tends toward a very flexible data model that is almost minimal in nature. It focuses on a modular primitive object design where the transfer structure is organized around the source data model in such as way that the transfer structure can be decomposed and reconstituted later as the target data model. This sort of transfer standard can be called a transfer mechanism. At the extreme such a transfer mechanism would have a very minimal data model that can be structured to accommodate an extremely wide range of source/target data models. As the range of the flexibility of such a transfer mechanism increases, the complexity of the mechanism would also increase as well. The goals here would to design a mechanism increase as well. The goals here would to design a mechanism that maximizes flexibility and minimized complexity. Since these goals conflict and are tradeoffs one with another, the task of designing a national transfer standard that serves the professional community efficiently is not an easy one.

DESIRED CHARACTERISTICS OF THE TRANSFER PROCESS

As one looks at this sort of spatial data transfer process, it is possible to identify basic characteristics that such a process should possess. The following discussion provides a listing and brief definition of the desirable characteristics that such a process should have. It is based on an earlier discussion by Moellering (1988) and has been adapted to the more general international situation. The discussion will be carried out in terms of general characteristics, cartographic objects, data quality, cartographic entities, and finally the transfer machinery.

Transfer Three Major Families of Spatial Data Structures

Currently there exist three major families of spatial data structures. A successful database exchange process should be able to transfer vector, raster and relational data structures and associated ancillary information.

Transfer Across Different Computer Architectures

For proprietary reasons there are many different kinds of internal machine architectures for computers. Many of these architectures are not compatible with each other. A good spatial database transfer mechanism will be able to transcend such architectural differences.

Transfer Databases Between Noncommunicating Parties

Typical database exchanges require prior communication, documentation and sometimes extensive discussion before any data can be transferred. The goal is to specify the standard such that this kind of communication is not necessary, but is built in as part of a standard data exchange mechanism.

Modular and Extendible

Because of the dynamic developments in the field of cartography and spatial data structures, the standard must be specified in a modular fashion. In this way new developments which need to be included in the future can be added in a modular fashion without disturbing other parts of the standard. Obsolete parts could be deleted in a similar way.

Robust and Reliable

Any database exchange mechanism must be specified such that it is robust and reliable. This means that it must not be sensitive to some data errors and occasional data corruption such that data perishability during transmission should be very low, thus insuring that the probability of correct transmission is quite high.

Media Independent
The exchange mechanism should be media independent such that spatial databases can be transferred by magnetic tape, direct telecommunication, hard or floppy disk, CD/ROM or any other appropriate medium.

Transfer Both Syntax and Semantics
A good transfer process with the above characteristics should be able to exchange the syntax, or structure of the data reasonably straightforwardly. However, the capture of the semantics, or meaning of the information transferred is much more difficult to insure. Transfer of the semantics is more difficult to do than is the transfer of syntax.

Facilitate the Transfer of Self-Describing Databases
Currently such databases are experimental or only a gleam in the designer's eye. However, it is clear that such databases are being developed and that they will be in common use within a decade. A standard must accommodate such exchanges.

Use Best Concepts and Existing Standards Available
Developing an efficient and comprehensive spatial database transfer standard in not a simple matter. Therefore one must be clearly aware of current research such that the best available concepts are used in the design of the standard. The items presented in the conceptual background form the crucial intellectual basis of a standard. Existing national and international standards that apply to the problem must be sought out and examined. Such existing standards should be utilized whenever possible.

Awareness of Work in cognate Areas
Every discipline such as cartography has interactions with its cognate areas, and therefore one must be aware of what is going on in those areas in the way of standards activities. In this case one must pay attention to four primary cognate areas: geographic and land information systems, photogrammetry and remote sensing, computer graphics and CAD/CAM, and, geodetic and land surveying.

Involve the User Community in the Process
This is a very important and time consuming activity. However, it is crucial because as the scientific solution to the standard is developed, a consensus must be developed in the professional community it is intended to serve.

Maintenance Authority
Developing cartographic standards is NOT a one time process. Since such a standard is being built in an area such as spatial data, there is active research taking place in many areas. One must realize that periodically the standard will have to be updated in the future. Therefore, it is clear

that the initial effort must be conducted such that future enhancements can be made without disturbing the existing parts of the standard. Hence, the notions of modularity, extendibility and robustness become very important so that the standard can be enhanced as new developments take place. All of this implies the founding of a maintenance authority to supervise the interpretation of the standard and to lead the work of actual maintenance. It is expected that the area with the highest demands on maintenance will be the cartographic entities.

CARTOGRAPHIC OBJECTS

Cartographic objects are the digital representation of entities and as such are the fundamental building blocks of any spatial data structure. Hence, they must be modular, and be capable of being assembled into more complex and higher dimensional structures.

Only Necessary to Define Primitive and Simple Objects
As with any standard, one must always clearly define the primitives in the system. Here it was also necessary to define some simple objects that are constructed out of primitives from lower dimensions. These simple objects serve as the primitive objects for the higher dimensions.

0-, 1-, 2-, and 3- Dimensional Objects Should be Defined
All four levels of dimensional primitives should be specified with provisions for building higher level objects if they become necessary. The standards currently do not explicitly define anything above 2-D primitives because not enough is known about them at the present time. 3-D and higher dimensional objects can be added as a future enhancement as they are better understood and needed by the community.

Geometry and Topology
When defining cartographic objects, it is necessary to define them so they include aspects of geometry and topology. How this is done can vary greatly depending the nature of the particular national standard being defined.

Coordinate Systems
For a spatial data standard objects must be validly defined for both Euclidean and simple curved surfaces such as the sphere and the ellipsoid.

Modular and Extendible Object Set
Given the above object requirements and general characteristics it is clear that the set of primitive and simple objects must be modular such that new ones, 3-D for example, could be added in the future, and so that compound and complex objects can build out of them.

DATA QUALITY

Data quality is an essential ingredient to any database transfer because the integrity of the data must be insured throughout the exchange process.

Truth in Labeling

The fundamental rubric under which the data quality portion of the standard operates is truth in labeling. This means that the producer of the database should be required to provide specified quality information for review by the prospective user. The user would review the quality report and decide whether that particular database is suitable for a particular application. It is anticipated that the better quality databases will see more frequent use than those of lesser quality and hence the specification of the quality report will create a tendency towards creating databases with improved data quality.

Flexible Quality Levels

In order to function as a national standard with broad application, it is clear that a national must have flexible quality levels in it. Fixed quality levels are product specific and therefore outside of a national standard, but may exist as internal quality specifications for particular products produced by individual organizations.

Various Levels of Testing Rigor

Because of the flexible quality requirements that are necessary, there are also different levels of testing rigor that could be applied in a particular situation. However, the report should explicitly state which level has been applied and what the results are.

CARTOGRAPHIC ENTITIES

Cartographic entities exist in the real world and are the things being represented modeled, analyzed and displayed in modern cartographic work.

Universal Set of Entity Definitions

The set of entity definitions for a national standard are desired to be universal and encompass all entities that may be coded in spatial databases at the national level. However there is a bottom limit of interest in detail for a national standard. A standard should be designed with conceptual hooks on the bottom to which additional entities and attributes could be attached.

Modular Entity Definitions

Entities are defined in a modular fashion with many aliases of analogous terms linked to them. This will serve to simplify the structural complexity of the definitions themselves, and pave the way for a nonheirarchical definition structure for the entities, if desired.

Extendible Set of Entities
Because of the modular definitions of the entities, a maintenance authority could update and modify the entity set as a regular and systematic part of its work.

Scale Independent These entity definitions are designed to be scale independent and not related to product specific scales. This is because a database may be used at a range of scales for various purposes, within reason.

Nonhierarchical Definition Structure
It is possible to define a set of entity definitions that are that are not hierarchically organized. Hierarchical entity definition structures are cumbersome to use, subject to judgmental error and are not very amenable to modern query retrieval systems. Therefore, an entity can be defined with many possible attributes and attribute values that could be assigned depending on the situation of the actual entity in the real world.

Designed for Modern Query Retrieval Systems
With the modular definitions, scale independence, and nonhierarchical definitional structure, entities, their associated attributes, and attribute values can be easily entered, retrieved and manipulated in modern database query management systems.

THE TRANSFER MACHINERY

The transfer machinery is what actually can convert the source data base to the transfer metafile form and later decode the metafile into the target database. The transfer metafile is the exchange file that is actually transferred between computer systems. This kind of transfer can occur with magnetic tape, by direct telecommunication, with hard or floppy disk, CD/ROM or almost any other kind of form because a standard should be media independent as pointed out earlier.

Transfer Three Families of Data and Ancillary Information
The data exchange mechanism should be designed to transfer data bases from all three of the fundamental data structure families: vector, raster, and relational, as well as information pertaining to attributes of the entities and associated quality information.
Modular Forms and Modules
Each of the three families of data structures should have a modular transfer form associated with it. Each of these three forms should be composed of a set of modules that transfer the primitives, attributes, quality information, and

other ancillary information. Every module of this kind is composed of a set of records that describe its own contents.

Preserve Syntax and Semantics
In the database transfer between systems, the syntax is preserved in the logical forms and modules. However, the semantics, or meaning, is implicitly preserved in the syntax of the exchange. Clearly, the syntax of the information is easier to preserve than the semantics.

Minimize External Documentation
A good database exchange process will reduce the need for external documentation, particularly if the software is written to run interactively. Ultimately, such a standard should be designed with the concept of being able to support self-describing databases in the future when they become available.

Generate Export Database Metafiles
The transfer converter processes the information from the source database and creates an exchange metafile for the actual database transfer between systems as shown in Figure 4. The transfer metafile will then be decoded into the target database by the converter.

Conform to Existing Standards
A particular standard must usually conform to existing national standards. In addition, International Standards Organization (ISO) standards should be used whenever practicable. These are the same or similar to existing standards used in other countries.

OVERVIEW OF STANDARDS EFFORTS

Currently there are about 20 ICA member nations and other international organizations working on the question of spatial database transfer standards. This work contains chapters that describe 18 of those efforts that are in progress. The range of these efforts includes those that are in the beginning stages of examining the problem to those that are nearly finished with a comprehensive effort. The following discussion summarizes each effort for which a written chapter has been received.

Australia
Work on a national topographic data transfer standard began in Australia in 1974, a very early effort. The effort focused on developing a transfer format for map data that was codified in 1981 as AS 2482 Interchange of Feature Coded Digital Mapping Data by Standards Australia. The standard was revised in 1984. In the later 1980's the group realized that a more comprehensive database transfer standard was needed. The groups are now evaluating their effort as well

as the work of several of the ICA national groups. Currently the U.S. Spatial Data Transfer Standard (SDTS) is a leading candidate for adoption. Further evaluation is proceeding.

Austria
Although digital map data processing has been going on for many years in Austria, it is only recently that an initial database transfer standard has been proposed by the Austrian Standards Institute. This standard is called ÖNORM A 2260, 1990 Datenschnittstelle für den Austausch geographisch-geometrischer Plandaten (Interface for digital exchange of geographic-geometrical data). This data transfer format is oriented primarily towards large scale maps and drawings. As such it currently geometry only with topology not explicitly specified. Since the standard was proposed in 1990, it has not yet been extensively field tested.

Canada
Canada is in the interesting position of developing more than one spatial database transfer standard, the first in the topographic area, and the second in the hydrographic area. The first effort began in the late 1970's by the Canadian Council of Surveying and Mapping (CCSM), now called the Canadian Council on Geomatics (COG). They have developed a Standard for the Exchange for the Exchange of Digital Topographic Data which was originally proposed in 1982. The proposed standard is a variation of a flexible format, and includes a data model, standard features and attributes, a quality specification, and a file format. Implementation and testing of the standard commenced in 1986. The standard is in its later stages of approval.

The second major effort in Canada is that of Map and Chart Data Interface Format (MACDIF) proposed by a group of agencies in the country. Work began in 1985 and is progressing. The MACDIF syntax is based on the Abstract Notation Syntax (ASN) and is intended "to standardize the interface point between two communicating entities regardless of the mapping or charting application".

Work is also going on to develop a Geographic Document Architecture (GDA). This effort is intended to operate is an Open Systems Interconnection (OSI) environment.

Peoples Republic of China
The challenge of cartographic database transfer is seen as a component of Geographic Information System (GIS) standardization in the Peoples Republic of China. Work is in the early stages of defining the problem in this context. It includes projects to develop land classification codes, creation of 1:4M geographic database, computerization of topographic symbols, among others.

Finland
Work on database transfer has been going on in Finland since the 1970's. In 1985 the Ministry of Agriculture and Forestry established a steering committee on the LIS project. The practical part of this work is being carried out by the National Land Survey. This work has resulted in National Administrative Standards (VHS) VHS 1041 on Geographical Data Representation and VHS 1040 on Message Description. These standards operate in an EDIFACT (ISO 9735, Electronic Data Interchange For Administration Commerce And Trade) environment. As such these data transfer standards seem to be a fairly flexible transfer mechanism. Additional work has been carried out on a Geographical Data Dictionary System (GDDS), a geographical query language that is a subset of SQL, as well as an EDIFACT interpreter.

France
Work on a national transfer standard in France was initiated by the National Council for Geographic Information (CNIG) in 1987. This led to a Commission for the Normalization of Exchange Formats. This group has been working on the database transfer standard in the areas of data model, data definition, exchange files, and exchange of messages. This has led to the adoption of the DIGEST data model (see DIGWG below) implemented on an ISO 8211 base that is called EDIGEO. Testing is in progress.

Federal Republic of Germany
The effort in the Federal Republic of Germany was begun in the early 1970's by the Working Committee of the Surveying and Mapping Institutes of the Länder (AdV). In 1982 the Uniform Database Interface (EDB) was published. This work has now become involved with the development of the Authoritative Topographic Kartographic Information System (ATKIS). As part of this work a data model, feature catalog and symbol catalog have been defined. As such EDBS appears to be a relatively fixed format. Current work is being carried out to define a more flexible version of EDBS in cooperation with other European efforts, notably CERCO.

Hungary
In the later part of the 1980's digital cartographic data processing became more widely spread and the need for a database transfer standard was realized. Initially efforts were made to review the Canadian, American, British and German standards proposals. Further work has resulted in a proposal for a Digital Cadastral Map standard (DCM). Work on this effort is continuing.

Japan
Although digital cartography has been used in Japan for many years, it is only relatively recently that standardization efforts have been actively pursued. In 1985 the Geographical Survey Institute formed the Committee for Digital Mapping

Standardization. This group defined the Standard Procedure and Data Format for Digital Mapping in 1988. This specification includes data quality, a feature coding system, and transfer format. As such this transfer standard is a somewhat flexible format based on fixed length records.

New Zealand
In 1987 Land Information New Zealand (LINZ) began closely evaluating possible alternatives for a spatial database transfer standard. After careful study, the U.S. Spatial Data Transfer Standard (SDTS) has become a leading candidate for adoption. However, LINZ is developing attribute definitions for use in the country. Further work is continuing.

Norway
In the late 1970's the Norwegian Ministry of Environment initiated the SOSI (Coordinated Approach to Spatial Information) project. From this project arose a defined transfer format, SOSI, in 1985. Object definitions and attribute codes were defined in 1987. The SOSI format is specified at four levels of use and as such is reasonably flexible. An ISO 8211 implementation vehicle is being considered. Currently SOSI only handles vector data, but future plans include handling raster and pictorial data.

South Africa
In 1985 a project team was formed at the CSIR Centre for Advanced Computing and Decision Support to develop a national spatial data exchange standard for South Africa. In 1987 SWISK 45 was published and has now become known as the National Exchange Standard (NES). NES provides for the transfer of vector and raster data. As such it is implemented by a relational transfer format that seems to be rather flexible. Current work on NES includes further enhancement of the standard and producing a user manual.

Sweden
In the early 1980's the Swedish Association of Local Authorities developed a simple standard, KF85, for the transfer of vector data. Since 1986 an effort has been under way by the Research & Development Council for Land Information Technology (ULI), to define a more comprehensive data transfer standard. Several Working Groups are carefully evaluating standards being developed in other countries. Currently, primary interest is focusing on an adaptation of the British NTF format on an ISO 8211 implementation that is compatible with the German ATKIS system in cooperation with CERCO.

Switzerland
The Swiss are currently developing a cadastral data system that will use a data transfer format called AVS (Amtliche Vermessungs Schnittstelle). They are also evaluating other

national efforts to develop such data exchange standards. There is interest in the possibility of adapting one of these standards as it applies to the situation in Switzerland. Evaluations are continuing.

United Kingdom

In 1983 the House of Lords Select Committee on Science and Technology recommended that a set of map data exchange standards should be established to replace earlier, more limited work. In 1986 the first version of the National Transfer Format (NTF) was issued. It is a reasonably flexible format that can transfer vector data with both geometry and topology. Raster and grid data can also be accommodated. NTF is specified on 5 levels of operation and transfers information on features and data quality. Current work involves evaluating the possibility of using ISO 8211 as the base implementation mechanism for NTF.

United States

The need to develop a comprehensive cartographic data transfer standard became evident in the early 1980's. To this end the ACSM National Committee for Digital Cartographic Data Standards was founded in 1982 under the sponsorship of the U.S. Geological Survey. Shortly thereafter the Federal Interagency Coordinating Committee for Digital Cartography was founded in 1983. Through their cooperative efforts The Proposed Standard of Digital Cartographic Data was published in 1988. Further work by the USGS Technical Review Board refined what is now known as the Spatial Data Transfer Standard (SDTS). SDTS is a very flexible exchange mechanism that will transfer vector, raster and relational data. Digital cartographic objects that have a wide variety of geometric and topological properties can be transferred. The standard consists of an object specification, entity specifications, data quality information, and the transfer mechanism. SDTS uses ISO 8211 as the basic implementation vehicle. SDTS has been extensively tested and is currently undergoing its final evaluation and polishing on its way to becoming a Federal Information Processing Standard (FIPS).

Comité Europeen des Responsable de la Cartographie Officielle

CERCO is an organization composed of the directors of the European national cartographic agencies. As such CERCO is interested in the possibility of defining a European Transfer Format (ETF). The leading contender seems to be a modification of the British NTF format adapted to an ISO 8211 implementation base. Discussions are continuing.

Digital Geographic Information Working Group

DGIWG is an international group with members of several European countries including the United States and Canada that was formed in 1983. Most of these countries are NATO members, although the DGIWG leadership has stated that this effort is independent of NATO. DGIWG has recently produced a proposal for a data transfer format called DIGEST, which is intended to handle raster, vector and matrix data. Digest is being amended to run on an ISO 8211 implementation base. A feature and Attribute Coding Catalog (FACC) has also been produced, along with several specifications for raster products. Field testing of DIGEST is continuing.

SUMMARY AND CONCLUSIONS

As the reader can see, there are currently many cartographic data-base transfer efforts going on in many parts of the world. The ICA Working Group on Cartographic Database Exchange Standards was founded in order to provide a forum for discussing this work at the international level.

In order to really understand the data transfer process, it is very important to understand the key concepts involved, such as real and virtual maps, their transformations, deep and surface cartographic structure, and the six Nyerges data levels. With those basic concepts in hand, one can begin to more fully understand the database transfer process. It is clear that the transfer metafile approach has advantages over the conventional pairwise converter approach because of the dramatically fewer number of data converters that would be necessary to specify, build and maintain.

It can be seen that it is possible to design and build a variety of cartographic database transfer processes that range from a simple fixed format that has rather limited use, to a more flexible format that has a much wider scope of use, to an exchange mechanism that can transfer most data organizations within the three data structure families in an environment that is hardware independent. These characteristics, plus many more discussed above provide the specifications for an ideal spatial data transfer mechanism.

Finally, one can review the large number of national efforts to develop such standards as discussed in the following chapters of this monograph. It can be seen that there is a very wide range of activities going on from some groups that are evaluating the situation to groups that have been working on the problem for many years and are in the final stages of a major transfer standard effort.

The question can also be raised concerning the future of this work, and the Working Group specifically. Clearly there is much more work to be done to continue dialogue between these various groups as the work continues. It is also possible to begin work to make comparisons between the standards being developed and proposed. It is possible that commonalities between the various transfer standards can be identified with the possibility of developing interlinkages between some of them. The question of a world standard has been raised, but this kind of a goal seems very elusive indeed. It seems possible that some of the standards could be adopted on a regional basis for common benefit. However, much more effort is required before this question can be efficiently answered.

REFERENCES

Moellering, H., 1980, "Strategies of Real-Time Cartography", Cartographic Journal, 17(1), pp. 12 - 15.

____________, 1983, Designing Interactive Cartographic Systems Using the Concepts of Real and Virtual Maps", in Wellar (ed.), Vol. II, Proc. of the Sixth International Symposium on Automated Cartography, pp. 53 - 64.

____________, 1984, "Real Maps, Virtual Maps and Interactive Cartography", in G. Gaile and C. Wilmott, (eds.), Spatial Statistics and Models, Dordrecht: Reidel Publishing Co., pp. 109-131.

____________, 1987, "Understanding Modern Cartography Using the Concepts of Real and Virtual Maps", Prod. of the 13th International Cartographic Conference, Morelia, Mexico, Vol. 4, pp. 43 - 51.

____________, 1988, "Fundamental Concepts that Form the Basis of the Proposed American Cartographic Data Exchange Standard", Cartography, 17(2), pp. 9 - 14.

____________, 1991a, "Whither Analytical Cartography", Cartography and Geographic Information Systems, 18(1), in press.

____________, 1991b, "Research Issues Relating to the Development of Cartographic Database Transfer Standards", in, J-C. Müller (ed.), Cartographic Research Agenda in the 1990's, Elsevier, forthcoming.

Morrison, J.L., 1974, "Changing Philosophical-Technical Aspects of hematic Cartography", American Cartographer, 1 (1), pp. 5 -14.

Nyerges, T.L., 1980, Modeling the Structure of Cartographic Information for Query Processing, unpublished Ph. D. dissertation, Dept. of Geography, Ohio State University, 203 pp.

____________, 1991, "Analytical Map Use", Cartography and Geographic Information Systems, 18(1), in press.

AUSTRALIAN STANDARDS FOR SPATIAL DATA TRANSFER

Andrew L. Clarke
Australian Surveying and Land Information Group
PO Box 2 Belconnen, ACT 2616, Australia

Standards Australia Subcommittee IT/4/2
PO Box 458 North Sydney, NSW 2059, Australia

ABSTRACT

The current Australian Standard for spatial data transfer, AS 2482 'Interchange of Feature-Coded Digital Mapping Data', is now ten years old and is unsuitable for modern spatial database and geographic information system applications. Standards Australia proposes to clone the U.S. 'Spatial Data Transfer Standard', with appropriate modifications for Australia, to supersede AS 2482.

INTRODUCTION

The Need for a Standard

One of the key economic benefits of GIS technology arises from the ability it provides to share spatial data among users. Data sharing reduces costs by avoiding duplication of data capture and maintenance. However, realisation of this benefit is dependent on the wide availability of an efficient and effective method for transferring spatial data between agencies and systems with different GIS hardware and software.

The Australian spatial data community is comprised of government, private and academic sectors. The government sector includes numerous agencies which utilise GIS in the Commonwealth Government, numerous agencies in each of the eight state and territory governments, and the larger of the nearly 500 local governments. The diversity and growth of the Australian spatial data community makes data transfer a critical issue.

The history and problems of the current Australian Standard for spatial data transfer, AS 2482, are described in this Chapter. The rationale and implications of the Standards Australia proposal to adapt the U.S. Spatial Data Transfer Standard as a new Australian Standard are also outlined.

Standards Australia
Standards Australia (formerly the Standards Association of Australia) is the national organisation for the promotion of standardization in Australia. It is an independent non-profit organisation administered by a Council comprising representatives from government, industry, professional groups and the community. Standards Australia is the Australian member of the International Organisation for Standards (ISO).

Standards Australia has published over 4000 Australian Standards on a diverse range of topics, and has about 1600 technical committees which prepare draft standards. Each committee is formed on a national basis with a balanced representation from all interested sectors of the relevant industry. New standards are initiated by authoritative sources external to Standards Australia, such as industry associations and professional societies. Technical committees either draft a new standard or adapt the work of the external body to the required format. Drafts are circulated for public comment and consensus must be reached within a committee before a Standard can be published. Australian Standards are not compulsory per se, but they are frequently referenced in statutory regulations and contracts making their use mandatory in specified situations.

The Committee relevant to spatial data is Information Technology Committee Number Four (IT/4), Geographical Information Systems. IT/4 has representatives from government agencies, academia, industry associations, professional societies and research bodies involved in surveying, mapping and land information. Development of spatial data transfer standards is the responsibility of Subcommittee IT/4/2, Geographic Data Exchange Formats.

AUSTRALIAN STANDARD 2482

Development
The development of the current Australian Standard for spatial data transfer, AS 2482, is summarised in the following chronology of events.

1974: The National Mapping Council (NMC) formed a Working Party to develop a standard for the exchange of digital topographic information. A key factor was the increasing use of private sector consultants to produce digital mapping data for government mapping authorities. The resulting 'NMC Standard on Exchange of Topographic Information on Magnetic Tape' was completed in 1978.

1979: The NMC submitted its Standard to the Standards Association of Australia (SAA, now called Standards Australia) for consideration as an Australian Standard. SAA constituted a more broadly-based committee to develop an Australian Standard based on the NMC work.

1981: The SAA published AS 2482-1981 'Interchange of Feature Coded Digital Mapping Data'. The Australian Standard was substantially different to the 1979 NMC Standard.

1982: A NMC Working Party developed a subset of AS 2482 'Recommended Procedures for the Interchange of Digital Mapping and Charting Data on Magnetic Tape'. The subset defined preferred options in places where the Standard allowed for alternatives.

1984: The SAA published a revised version of the Standard AS 2482-1984. This was an extension of the 1981 Standard, adding more feature codes and improving the scope and content of various record types.

1985: The NMC revised its recommended procedures document to reflect AS 2482-1984 and the experience of members in the use of the Standard. The NMC document was titled 'Recommended Procedures for the Interchange of Small and Medium Scale Digital Vector Topographic Mapping Data' to reflect the NMC view of the narrow scope of AS 2482.

1987: The SAA formed Committee IT/4, Geographical Information Systems. This was partly in response to a request from the NMC to change the title of AS 2482 to reflect its narrow application. Key outcomes from the initial meeting of IT/4 were to produce a third version of AS 2482 by incorporating the NMC subset, and to assess the U.S. Draft Standard for Digital Cartographic Data with a view to adopting it as a basis for development of a new Australian Standard to supersede AS 2482. Subcommittee IT/4/2, Geographic Data Exchange Formats, was formed to undertake these tasks.

1989: Standards Australia published AS 2482-1989 (Standards Australia 1989). AS 2482-1989 is compatible with AS 2482-1984 and includes an Appendix based on the 1985 NMC subset. The description of the scope was changed and other minor changes were made to reflect developments relating to the Australian Geodetic System and to enable identification of versions of AS 2482.

AS 2482 has been a moderately successful Standard. It is widely used by government mapping agencies who acquire data from the private sector, and who distribute data to users.

Concepts

AS 2482 specifies a file and record structure for the interchange of point and vector digital mapping data. It is not intended to be used for the transfer of polygon, raster or topologically structured spatial data, nor for attribute data which may be associated with the spatial data. It is designed for interchange on magnetic tape and makes use of existing national and international standards for tape labelling and encoding. A hierarchical system of four-digit feature codes defines about 750 cultural, hydrographic, relief and vegetation features. Users may also define four-digit feature modifiers to further specify map features.

The general structure of an AS 2482 map data file, in accordance with the NMC subset, is as follows:

Tape Label: Fixed length header with tape identification information.

File Headers: Two fixed length headers containing basic file identification, the creation date, and format information.

Essential Information Record: Fixed length record defining coordinate systems, scale factors and offsets.

(Basic) Descriptive Information Records: Six fixed length records defining: the map number, name, scale and theme; the owner, agency and contact person; the source, source scale and source date; the date digitized and date last revised; the estimated root mean square error in X, Y and Z; and the camera focal length and flying height (for digital photogrammetric data).

(Other) Descriptive Information Records: Fixed length records containing other descriptive information, if required, such as non-standard feature codes, feature modifiers, or donor-defined coordinate systems.

Feature Records: Variable length records for each feature. Each record has two or three segments: Header Segment, defining the record length, nature of feature (point or line), feature code and modifier, and number of axes (Z, XY, or XYZ); Detail Segment for Line, Point or Text Data, containing the feature coordinate values; and if required a Detail Segment for Identification/Name, containing textual data such as the feature name.

End of File: Two fixed length labels defining the end of the file.

Problems

AS 2482 represents the state-of-the-art in the late 1970s for computer-assisted map production. The technology then comprised data acquisition through digital photogrammetry or table digitizing, followed by production of map reproduction material on precision vector plotters.

Initial criticisms of AS 2482 were that the options provided in various parts of the Standard made it difficult for users to write comprehensive and robust transfer software, and that the specified feature codes did not satisfy large-scale mapping applications. The NMC subset partially addressed these criticisms. However, with the development of large spatial databases and analytical applications of spatial data, based on GIS technology, AS 2482 was also seen to have some serious conceptual problems. These include:

- o does not support polygon, grid or raster data types;
- o does not support topologically structured data;
- o has minimal provision for data quality information;
- o has minimal provision for attribute data.

These problems merely reflect the original purpose of the Standard, which was to facilitate data transfer for digital topographic map production. It is therefore not a criticism of those involved with its development to say that it is not suitable for use as a general-purpose spatial data transfer standard for GIS and related applications.

U.S. SPATIAL DATA TRANSFER STANDARD

The U.S. Spatial Data Transfer Standard (SDTS) is described in the U.S. chapter of this monograph. Only a brief summary is provided in this chapter.

Development

The development of the U.S. SDTS commenced in 1980, with the final draft being submitted to the U.S. National Institute of Standards and Technology (NIST) in 1990. A feature of the development process has been the extensive consultation and testing.

After approval by NIST for the SDTS to become a Federal Information Processing Standard (FIPS) it will be submitted to the American National Standards Institute for promotion as an ANSI Standard.

Concepts

The three parts of the SDTS are outlined below.

Model, Specification and Quality: Part 1 provides a general model for spatial data, a transfer specification, and a specification for data quality reporting. The data model comprises entities, attributes and objects and is based on the concepts of phenomenon, classification, aggregation, generalization and association. The transferspecification provides modules for global information, for attribute data, for vector, raster and composite objects, for graphic representations and for data quality information. The quality specification utilises a 'truth in labelling' approach, requiring users to report what is known about the lineage, positional accuracy, attribute accuracy, logical consistency and completeness of the data.

Cartographic Features: Part 2 provides a non-hierarchical and extendible model for a spatial data dictionary, comprising entity, attribute and attribute value definitions. Some initial definitions are given (for topographic and hydrographic features) and more will be developed by the maintenance authority. Users may supply their own entity and attribute definitions within the transfer set.

Transfer Mechanism: Part 3 defines the transfer mechanism, which is implemented in an existing general-purpose interchange standard ISO 8211 'Information Processing Specification for a Data Descriptive File for Information Interchange'.

A NEW AUSTRALIAN STANDARD

Proposal

Two approaches were available to Standards Australia for the development of a new standard to supersede AS 2482: either start from scratch and write a new standard in consultation with the Australian spatial data community, or adapt an existing standard to suit the Australian requirements. The first approach would involve many years of effort by many people and could only be justified if no suitable existing or proposed standards could be identified.

Many existing and proposed standards for spatial data transfer are described in other chapters of this publication. While each may have advantages, the proposed U.S. SDTS was considered to be the most appropriate. It overcomes the conceptual problems of AS 2482, it has been developed with extensive user consultation, it will be supported by the major North American GIS vendors who are active in the Australian GIS market, and implementation by Australian GIS vendors will assist those vendors in penetrating the U.S. GIS market. Further, it is considered that the U.S. SDTS is more likely than others to be adopted by Australia's Asian and Pacific neighbours.

Technical benefits of adapting the U.S. standard include:

- It will be applicable to most of the spatial data community, particularly GIS, LIS, remote sensing and computer-assisted cartography users.
- It will enable transfer of all spatial data types (topologically structured and unstructured vector data, raster data) and the associated attribute data.
- It will assist all levels of communication between spatial data users through definition of a general spatial data model.
- It provides a structure for data quality reporting.
- It provides a structure for the development and

maintenance of Australian entity and attribute definitions.

Standards Australia therefore proposes to clone the U.S. standard when it is published as a FIPS, with the minimum necessary modifications to make it suitable for Australian use. Public consultation will be on the question of cloning rather than on the detail of the standard.

Adaption
Three areas of modification to adapt the U.S. SDTS to Australia have been identified: referenced standards, coordinate systems, and entity and attribute definitions.

Some of the existing standards referenced in the U.S. SDTS may not be applicable or valid within Australia. Alternative standards may need to be substituted, or the referenced standards may be adopted for Australia or incorporated within the new Australian standard. ISO 8211 has already been cloned as AS 3654-1989.

The Australian version must refer to the Australian Map Grid, the Australian Height Datum and to other relevant coordinate systems. No problems are envisaged with this modification.

The U.S. definitions for topographic and hydrographic features are not generally applicable to Australia. Australian definitions for these and other types of geographic entities and attributes will be required, in accordance with the model structure included in the U.S. SDTS. Subcommittee IT/4/4, Entity and Attribute Definitions, has been formed by Standards Australia to coordinate this work. Working groups are being formed for the following data types:

- topographic and hydrographic;
- geological and geophysical;
- land use;
- natural resources;
- cadastral;
- street addressing;
- utilities.

The existing draft standards and coordinating mechanisms of groups such as the Australian Land Information Council and the Inter-Governmental Advisory Committee on Surveying and Mapping will be utilised in the development of the Australian definitions. The Australian Surveying and Land Information Group has produced a set of test files of topographic, census, remote sensing, cadastral and utilities data to facilitate testing of the standard and validation of transfer software.

Assuming that the U.S. SDTS becomes a FIPS by the end of 1990 and that there is consensus on the question of cloning, the new Australian standard should be available by early 1991. AS 2482-1989 would remain as a valid Australian Standard for an overlap period of some years. Ongoing management of the new standard by Standards Australia will include continuing development and maintenance of entity and attribute definitions, implementation of revisions made by the U.S. maintenance authority, and promotion of the standard to the Australian spatial data community.

Implications
A number of implications arise from the proposal to clone the U.S. SDTS. These include:

- There is a need for Australian testing of the standard, both to validate its applicability to Australian data types and to develop local expertise in its concepts and implementation.

- There may be a need for an Australian support group to validate transfer software and to provide training, documentation and support to users. Such a support group may also take responsibility for maintaining the entity and attribute definitions database. Standards Australia does not have the facilities to offer these services, but would of course cooperate in its operation.

- The spatial data transfer specification within the standard is complex, reflecting the complexity of structured spatial data. Software development by users may not be practicable, so there is a need to encourage local spatial software vendors to support the standard.

- The overheads in creating a conforming set of files may inhibit use of the standard for small data volumes, on-line transfers, and transfers involving primarily attribute data.

- The standard will not be applicable to all agencies and all spatial data types. Defence agencies have international obligations which include support of alternative standards. Some industry sectors may consider that the effort required to conform with a general-purpose spatial data transfer standard exceeds the benefits for their specialist applications.

- Comprehensive entity and attribute definitions must be developed by the user community, within the framework provided by the standard and Subcommittee IT/4/4. This will be a major task but it is essential if the full benefits of standardisation are to be realised.

- Full compliance with the data quality report will be challenging, but should yield benefits for both data producers and users.

- Promotion of the standard will require a concerted effort from Standards Australia, agencies that produce and distribute spatial data, and agencies that receive spatial data.

CONCLUSIONS

Australia was ahead of many countries in the adoption of a national standard for spatial data transfer when AS 2482 was first released in 1981. However, AS 2482 is not suitable for GIS applications and a new general-purpose spatial data transfer standard is urgently required. The Standards Australia proposal to clone the U.S. SDTS offers significant economic and technical benefits. The Australian spatial data community is now addressing the implications of this proposal.

REFERENCES

Standards Australia, 1989, Australian Standard 2482-1989 Geographic Information Systems - Geographic Data - Interchange of Feature-Coded Digital Mapping Data, Sydney.

THE AUSTRIAN STANDARD FOR DIGITAL EXCHANGE OF GEOGRAPHIC-GEOMETRICAL DATA

Wolfgang Kainz
Department of Geography, University of Vienna
Universitätsstraße 7, A-1010 Vienna, Austria
Electronic mail (bitnet): A6262DAB@AWIUNI11

ABSTRACT

Initiated by the pressing needs of the utility companies to be able to exchange digitized map data the Austrian Standards Institute developed a standard for the exchange of geographic-geometrical data. This standard is an exchange mechanism which clearly defines the procedures for the mapping of data from a sending GIS to an interface and from the interface to the receiving GIS. The interface file contains objects with graphic and non-graphic attributes. At the moment the standard does not transfer topology. Feature codes are also not included. Since the standard was introduced only recently, testing results cannot be reported.

INTRODUCTION

Digital processing of spatial data has a long tradition in Austria. During the seventies the first applications of digital mapping were done mainly in governmental agencies showing demographic data in form of choropleth maps based on political boundaries. However, graphic hardware was expensive and software was hardly available except for simple mapping tasks.

With the advent of affordable mini- and microcomputers the interest in digital mapping grew. Several research and development organizations began to use digital cartographic data and GIS. Among those were the Institute for Image Processing and Computer Graphics of the Research Center Joanneum in Graz, the Austrian Institute for Regional Planning in Vienna and the Austrian Research Center in Seibersdorf. Especially the Institute for Image Processing and Computer Graphics concentrated on the development and application of GIS and remote sensing software and the implications of spatial data processing (Buchroithner et al., 1985; Kainz, 1986). The universities gradually introduced GIS methods in their geography and surveying curricula, and today most of the major universities offer GIS and digital cartography courses.

The use of digital cartographic data in the public sector lead to the installation of geographical information systems in almost all of the nine provinces of Austria. They are set up and maintained by the planning agencies of the provincial governments. In recent years we have seen a tremendous growth of digital spatial data sets in the fields of environmental protection and monitoring, regional and urban planning and the conservation of cultural goods. The Federal Environmental Agency is collecting and exchanging data with the governments of the provinces, which makes the need for compatible exchange formats understandable.

The Federal Office for Metrology and Surveying is responsible for the topographic and cadastral mapping of the country. Several steps have been taken to provide digital cartographic base data for the customers. Today the political boundaries, the positions of the main geodetic points and a high resolution terrain model are available in digital form. The digitization of the cadastral maps in the scale 1:1,000 is in progress.

BACKGROUND AND HISTORY

Since almost all of the provincial planning agencies in Austria use a GIS software from one vendor, the problems with the exchange of digital data are not severe. The situation is completely different, however, with the utilities. They use many different systems, and the relatively high costs for data capture have lead to efforts towards a standardization of a data exchange mechanism.

The Austrian Post and Telecommunication Administration, the City of Vienna and other big agencies spend enormous amounts of money for surveying and digitizing the network of power, communication or sewer lines. Different data interfaces of the systems in use impede the exchange of data and increase the costs.

As a consequence and to overcome these problems a subcommittee in the Austrian Standards Institute was formed by the end of 1988. The subcommittee AG084a.01, under the chairmanship of H. Plach of the Technical University of Vienna, was to work out an exchange mechanism for digital geographic-geometrical data. Their primary intent was not to develop a standard for all kinds of cartographic data, but mainly to provide a fast and flexible solution for the pending problems with the exchange of facility management data.

Fifteen individuals from various fields of application met in more than 20 sessions between the end of 1988 and March 1990. Their work was completed in March 1990 and the draft standard submitted for approval. With June 1, 1990 ÖNORM A 2260 became an official Austrian standard under the name Datenschnittstelle für den Austausch geographisch-geometrischer Plandaten (interface for digital exchange of geographic-geometrical data) (ÖNORM A 2260, 1990). As stated in its German title the standard is particularly oriented towards large scale maps and drawings (the German term Plan stands for large scale map or drawing). For cartographic data it is of limited use.[1]

SHORT DESCRIPTION OF THE STANDARD

The standard provides a mechanism for the exchange of geographic-geometrical map data between different geographical information systems. Fig. 1 shows the concept of the exchange mechanism and the domain of the standard. It should be noted that this standard not only defines the structure of the exchange file, but also standardizes the mechanism of data exchange, the mapping to and from the interface.

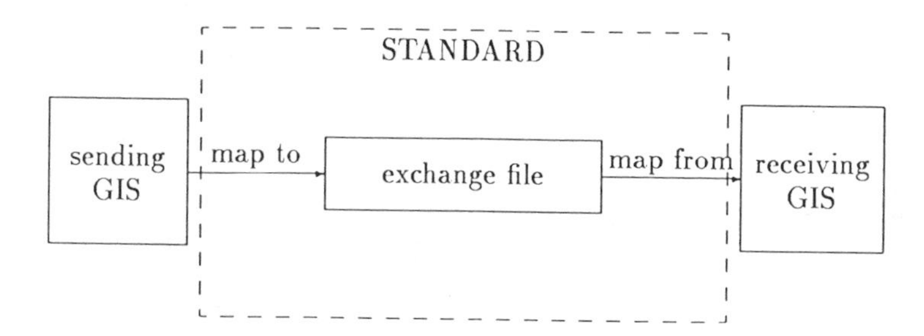

Figure 1. Concept of the exchange mechanism

The standard is primarily oriented towards large scale utility maps and is based on a common coordinate system. For Austria this is the Gauß-Krüger system with three zones and central meridians at 10 20', 13 20', and 16 20' east of Greenwich. In certain cases other coordinate systems are allowed.

---

[1]In the chapter on the scope of the standard it reads (translation by the author) ... This standard is only of limited application for the transfer of cartographic data. ... (ÖNORM A 2260, 1990, Section 1, page 2).

The exchange file is sequential and contains only printable characters of either the ISO 7-bit coded character set (ISO 646, 1983) or the ISO 8-bit single-byte coded graphic character sets, Part 1, Latin alphabet No. 1 (ISO 8859-1, 1987). Every record has an identifier according to the exchange mechanism. The interface has a hierarchical object-oriented design, the overall structure of the exchange file is shown in Fig. 2.

Each data set may transfer one or more maps, however, the coordinates must be unique for the whole set. If there are coordinates for each map in a multi-map set, the map sheets must not overlap.

Header
Map$_1$
Object-Class$_1$ (*attributes*)
Object$_1$ (*attributes*)
graphic elements (*attributes*)
⋮
⋮
Object$_n$ (*attributes*)
graphic elements (*attributes*)
⋮
⋮
Object-Class$_n$ (*attributes*)
Object$_1$ (*attributes*)
graphic elements (*attributes*)
⋮
⋮
Object$_n$ (*attributes*)
graphic elements (*attributes*)
⋮
<list of coordinates for map$_1$ >
Map$_1$ Trailer
Map$_2$
⋮
Map$_n$
⋮
<<list of coordinates for the whole set>>
Trailer

Figure 2. Structure of the interface file (after ÖNORM A 2260 (1990))

The set header contains general information regarding the sender of the data set and the coordinate system used for transfer. At least one map must follow the header.

According to the standard a map is a mapping of a subset of a geographical information system to the interface. The map header holds the name, the extent and the creation date of the map.

Each map contains object classes, every class must have at least one object which consist of points, lines, areas and texts, called graphic elements. Every object has an identifier and an object code. Any item in the hierarchy from object classes down to graphic elements can have graphic and non-graphic attributes. For object classes and objects the attri-butes are optional, graphic elements, however, must always have graphic attributes assigned, such as symbol number, line or hatch style, and text font. Non-graphic attributes are optional.

The list of coordinates contains all coordinates of the map. They are identified by a unique index, which is referenced by the graphic elements. Any point exists only once for a map and may have information assigned concerning point type, accuracy and lineage. Text reference points are added after all coordinates. They mark the start point (lower left) of a text string on the map.

The map trailer contains statistical information for the map, such as the total number of object classes, objects, coordinates and text reference points. The set trailer marks the end of the data set and holds totals for all maps.

Comments may be added to the data stream at any location. They are used to transfer additional information, e.g., spline algorithms or specifications of coordinate systems.

The standard also defines procedures for the mapping to and from the interface, the exchange and processing of code tables and splines, and the verification of codes and formats.

The mapping to the interface is defined as follows:

- specify the extent of the data set to be mapped
- choose object classes
- automatic choice of the respective graphic objects
- transformation of code tables of the sender to code tables of the receiver, and specification of format and contents of non-graphic attributes
- create exchange file

The mapping from the interface to the receiver's GIS is defined by the following steps:

- o read exchange file from media
- o list or print header and statistical information
- o check whether shipment is in order
- o copy all point and text reference coordinates to an intermediate file o convert code tables of the sender to code tables of the receiver, and arrange for proper processing of format and contents of non-graphic attributes
- o convert the data set to the GIS data format of the receiver

Code and attribute tables are not part of the exchange file. They must be transferred separately. Code conversion can be done either by the sender or by the receiver. The following tables are concerned:

- o symbol table
- o line table
- o hatch table
- o font table
- o color table
- o object codes

The full text of the standard can be obtained from the Austrian Standards Institute by writing to:

Österreichisches Normungsinstitut
Heinestraße 38, Postfach 130
A-1021 Vienna, Austria

FUTURE WORK AND DIRECTIONS

The main intent with the introduction of this standard was to have a formal, yet flexible, way of exchanging geographic-geometrical data. Therefore some of the features that have been standardized in other countries are not included in the Austrian standard. However, a standardization of object codes would be desirable. This is the reason why space for an 8-character object code is reserved in the exchange mechanism. This space can currently be used for the exchange of non-standard feature codes.

Since the standard has been introduced only recently, experiences with its application can not yet be reported.

The Austrian Post and Telecommunication Authority will, however, demand that all coordinate data surveyed and captured for their use has to be delivered according to the standard. It is anticipated that the major users of AM/FM systems in Austria will also require that all digital data has to be delivered according to the standard.

CONCLUSION

Austria is one of the first countries with an official standard for the exchange of geographic-geometrical data. It is designed to be a flexible mechanism for special fields of application. Due to this limited scope and perhaps also to the relatively short time it took for its development, there are some points that should be considered for future revisions of the standard.

Since all maps are based on spatial data, it would be highly desirable to extend the scope of the standard to all kinds of spatial data shown on today's analog and digital maps, including vector and raster representations. The extended version should also contain implicit and explicit topological relations between coordinate data.

ACKNOWLEDGEMENT

The author would like to thank H. Plach for his useful comments and his support during the compilation of this paper. The support by the Austrian Standards Institute with a copy of the standard's text is gratefully acknowledged.

The opinions expressed in this publication are those of the author and not necessarily those of the Austrian Standards Institute.

REFERENCES

Buchroithner, M., Kainz, W., Ranzinger, H., 1985, "Geowissenschaftliche Aktivitäten am Institut für Digitale Bildverarbeitung und Graphik der Forschungsgesellschaft Joanneum, Graz", Österreichische Zeitschrift für Vermessungswesen und Photogrammetrie, 73, Heft 1.

ISO 646, 1983, "Information processing; ISO 7-bit coded character set for information interchange", International Organization for Standardization.

ISO 8859-1, 1987, "Information processing; 8-bit single-byte coded graphic character sets, Part 1: Latin alphabet No. 1", International Organization for Standardization.

Kainz, W., 1986, "Some Ideas on the Standardization of Geoprocessing", Bulletin du comité français de cartographie, 109-110, pp. 95-97.

ONORM A 2260, 1990, "Datenschnittstelle für den Austauschgeographisch-geometrischer Plandaten", Österreichisches Normungsinstitut, Wien.

CANADIAN EFFORTS TO DEVELOP SPATIAL DATA EXCHANGE STANDARDS

Timothy V. Evangelatos
Chief, Cartographic Development,
Canadian Hydrographic Service,
Fisheries and Oceans

M. Mosaad Allam
A/Director GIS Division,
Surveys, Mapping & Remote Sensing Sector,
Energy, Mines and Resources

ABSTRACT

This chapter describes some of the efforts currently underway to satisfy the need for standards for the exchange of geographical data in Canada. With the exception of digital topographic data, for which standards have been implemented, the probability of one solution appears unlikely in the short term, as some of the mapping and charting agencies have made or will be making commitments before acceptable national standards are available. Five related activities are described.

INTRODUCTION

Canada's large size, the diversity of its political structure and its two official languages may make it difficult to reach a broad consensus on national standards for the exchange of geographical information. Nevertheless, there are currently five related activities that are attempting to provide the basis for one or more solutions. These are described in this chapter in the following chronological order of development:

Digital Topographic Data Standards
This activity, organized under the auspices of the Canadian Council on Surveying and Mapping (CCSM), has been on-going for more than a decade and has successfully developed an exchange format which has been implemented by the federal government and several provinces. Standards for the classi-

fication and quality evaluation of digital topographic data have also been produced.

Map Data Interchange Format (MDIF)
Developed by the Ontario Ministry of Natural Resources, this work is expected to have a strong influence upon any final national exchange format that may be adopted in Canada. It uses the same model as the CCSM format but employed other international telecommunication standards for its implementation.

Map and Chart Interchange Format (MACDIF)
Similar to MDIF, but with a broader context to include special applications such as the exchange and updating requirements for nautical chart data. This included special needs to fully reproduce paper charts as well as to providing a mechanisn for the distribution of data for electronic charts.

Committee On Geomatics
Created under the auspices of the Canadian General Standards Board in 1989, this committee will endeavour to establish a broad range of spatial standards. Five working groups operating under the direction of the main committee may develop unique standards, or if it is appropriate, adopt standards developed by others.

Geographical Document Architecture (GDA)
The CCSM standards were developed for exchange by magnetic tape. MDIF and MACDIF were developed for telecommunications. Neither solution is fully adequate and a compatible exchange mechanism for both physical and electronic media is required. GDA is an attempt to fill this need.

The specifications of the standards or relevant techni-cal reports for each activity is summarized at the end of each section with the name and address of the person who can provide the material.

CANADIAN COUNCIL ON SURVEYING AND MAPPING (CCSM) NATIONAL STANDARDS FOR THE EXCHANGE OF DIGITAL TOPOGRAPHIC DATA

During the past two decades, mapping organizations have witnessed a tremendous growth in digital mapping systems. Consequently, the mapping community has seen a proliferation of various approaches to the acquisition, coding and storage of digital topographic and cartographic data. As a result, there are as many data bases, with specific file structures, as there are systems.

The continuing proliferation of these varying approaches tends to hinder the orderly development of national standards for digital topographic data. In May 1978, a Government Task Force on National Surveying and Mapping recommended that the

Surveys and Mapping Branch, Energy, Mines and Resources Canada (EMR) develop national standards for digital topographic data exchange.

In October 1978, under the auspices of the Canadian Council on Surveying and Mapping a steering committee comprised of representatives of the federal and provincial governments, Canadian universities and the private sector was set up to establish a mandate and to define the tasks reuired for the development of the standards by the technical committees. A permanent secretariat was also established in EMR to provide continuity and to assist in the development of the standards.

The CCSM National Standards have been developed by three technical committees and one subcommittee over the past eight years in four phases:

Phase 1 started in 1979 with the formation of three technical committees. The committees were responsible for:

- o the classification and coding of digital topographic data;
- o the quality evaluation of digital topographic data; and
- o the Electronic Data Processing (EDP) file exchange format.

This phase ended with the publication of the first draft of the standards (CCSM, 1982) and its distribution to more than 700 organizations and individuals in Canada and abroad.

Phase 2 started in 1982 and dealt with the revision of the standards as a result of the comments received from the recipients of the first draft and the initial testing of the standards by federal and provincial mapping organizations. This phase ended with the publication of the second draft (CCSM, 1984a) and the adoption of the standards for topographic feature classification/coding and the standards for quality evaluation, by the Canadian Council on Surveying and Mapping.

Phase 3 started in 1984 with the formation of a technical subcommittee for the development of:

- o A standard digital topographic information model (DTIM);
- o A revised or a new CCSM National EDP file exchange format.

The Digital Topographic Information Model report (CCSM, 1985) was published and this phase ended with the adoption of a new CCSM national EDP file exchange format (CCSM, 1986a).

Phase 4 started in 1986 with the implementation and testing of the CCSM National Standards. Software was developed and installed in several federal and provincial mapping agencies to test the standards.

Committee Memberships:
Members of the technical committees representing federal and provincial agencies, and the private sector were drawn from:

- o Energy, Mines and Resources Canada
- o Department of National Defence
- o Statistics Canada
- o Fisheries and Oceans Canada
- o Alberta Bureau of Surveys and Mapping
- o British Columbia Ministry of Environment
- o Land Registration and Information Service, New Brunswick
- o Ontario Ministry of Natural Resources
- o Ontario Ministry of Transportation and Communication
- o Ministère de l'Énergie et des Ressources du Québec
- o Saskatchewan Central Survey and Mapping Agency
- o Société de Cartographie du Québec
- o Manitoba Mines, Natural Resources and Environment
- o British Columbia Hydro
- o Canadian Association of Aerial Surveyors
- o Kenting Earth Sciences Limited
- o Shell Canada Resources Limited
- o Terra Surveys Limited

Digital Topographic Data Model
A data model was developed to define the data components that are required to describe digital topographic data, and to define the rules according to which data are to be structured. The data model defines the following basic information:

- o Spatial data types, line and area;
- o Feature classification code as defined in the CCSM

coding standards (CCSM, 1984b);

- o Unique identification number (ID);

- o Spatial locational data: a point is represented by a single coordinate triplet (X,Y,Z); a line by two or more coordinate triplets; and an area is defined spatially by the lines that bound it;

- o User defined attributes;

- o Spatial relationships (topology): the minimum allowable relationships are connectivity and adjacency.

A detailed description of the data model is given in the Standards for a Digital Topographic Information Model (DTIM) (CCSM, 1985).

CCSM File Format
The basic components of topographic data to be represented in the CCSM file format were given in detail in the DTIM report (CCSM,1985) The following basic requirements were considered:

- o Machine and language independence - to allow for the exchange between difference computer systems;

- o Self-definition - having a standard internal format wherein structures are defined at the start of the data sets and/or data subsets which describe the remainder of the data;

- o Modularity and expandibility - to handle new types of data;

- o Flexibility - to allow for options in data components, i.e., giving the user options for the transfer of attributes and/or topology or no-attribute and/or no-topology

Three levels of issues were addressed in the CCSM file format:

- o Superstructure - where and how each type of file is placed on the physical medium;

- o File structure - where and how records are placed within each type of file; and

- o Data structure - the format and inter-record linkages within each record type.

The CCSM format specifies that entities (spatial and attribute information) be combined into data themes, which are then combined topologically into data groups which are in turn combined geographically into data sets, and are then finally combined into a logical volume.

A logical volume contains one or more Volume Directory Files (one for each physical volume), one or more Data Set Files (one for each data set) and any number of Data Theme Files for each data set. Each Data Theme File contains one Data Theme Header Record, and any number of Attribute Descriptor Records and Entity Records.

A detailed description of the CCSM file format is given in the Standard EDP File Exchange Format for Digital Topographic Data (CCSM,1986a).

Classification/Coding of the Features
For the classification of topographic features, a three-structured approach consisting of four levels was adopted as follows: (I) Class, (II) Category, (III) Feature and (IV) Attribute. In level I the following major classes were identified: Designated Area, Building, Structure, Roadways and Railway, Utility, Delimeter, Hydrography, Hypsography and Land Cover. Level II represents the category of information, which is a breakdown of the Class level. For example, under Building there are categories such as commercial, governmental, residential, etc. Topographic features were included on level III and essential attributes on level IV.

For the coding, alpha characters were used for the Class and Category level, while five-digit integer codes were used for the features. The essential attributes in level IV, which form an integral part of the feature were allowed a three-digit integer code. Detailed information on the classification and coding is contained in Part I, of Volume I of the CCSM standards (CCSM, 1984a), and a complete listing of all the topographic features codes is included in Volume II (CCSM, 1984b).

In addition, Volume II includes a dictionary of the terms which define all the classified and coded features.

Quality Evaluation of Digital Topographic Data
The standards include a set of rules by which producers of digital topographic data are able to calculate realistic estimates of the accuracy of the data contained in their files.

Other issues addressed in these standards were: the definition of all the parameters required for the management, processing, and retrieval of the data. The data quality standards also included a review of the important mapping accuracy specifications and of digital data acquisition techniques and technologies. A detailed account of these standards is included in Part II, Volume I (CCSM 1984a).

Evaluation of the CCSM National Standards
In their report, the Digital Topographic Information Model Working Group (CCSM, 1986b) recommended that the CCSM National Standards for the Exchange of Digital Topographic Data should be tested to measure their success in meeting the users/producers requirements.

Based on this recommendation, the CCSM, at the 1986 annual meeting, resolved that:

- The EDP file format proposed by the technical committee on EDP Standards be accepted as an interim standard;
- An operational and economic evaluation of the CCSM Standards take place through pilot projects and that such projects be designed to test the various elements of the Standards, i.e., as a minimum, the feature codes, data quality, EDP file headers, data evaluation and standard verification;
- A permanent Secretariat be established in the Surveys and Mapping Branch, EMR, where some of its responsibilities would be to coordinate the pilot projects, to update and distribute publications and to publicize the interim Standards within the surveying and mapping community and related fields.

In order to implement resolution (b), the Canada Centre for Geomatics, Surveys, Mapping and Remote Sensing Sector, EMR developed the necessary EDP file exchange software which would facilitate the conversion to and from the CCSM format through interfacing subroutines.

This software was then implemented at various federal and provincial agencies in an attempt to promote the Standards and to demonstrate their capabilities. The status of this effort is shown in table 1 (Piche, 1988).

The Alberta Bureau of Surveying and Mapping (ABSM) conducted a thorough evaluation of the file format to test the conversion process to and from CCSM of their digital map data at the scales of 1:1000, 1:5000, 1:20,000, 1:250,000 and 1:1,000,000. The test proved that this was possible but pointed to the need for the coding of textual information and parametric arcs. The recommendations made by ABSM and other organizations were implemented and version 1.2 of the CCSM File Format was produced.

In addition, the Topographic Mapping Division, EMR, set a policy for fiscal year 1990/91 whereby all data disseminated or acquired for inclusion in the National Topographic Data Base (NTDB) have to adhere to the CCSM Standards.

In their report, the Digital Topographic Information Model Working Group (CCSM, 1986b) recommended that the CCSM National Standards for the Exchange of Digital Topographic Data should be tested to measure their success in meeting the users/producers requirements.

Based on this recommendation, the CCSM, at the 1986 annual meeting, resolved that:

- o The EDP file format proposed by the technical committee on EDP Standards be accepted as an interim standard;
- o An operational and economic evaluation of the CCSM Standards take place through pilot projects and that such projects be designed to test the various elements of the Standards, i.e., as a minimum, the feature codes, data quality, EDP file headers, data evaluation and standard verification;
- o A permanent Secretariat be established in the Surveys and Mapping Branch, EMR, where some of its responsibilities would be to coordinate the pilot projects, to update and distribute publications and to publicize the interim Standards within the surveying and mapping community and related fields.

In order to implement resolution (b), the Canada Centre for Geomatics, Surveys, Mapping and Remote Sensing Sector, EMR developed the necessary EDP file exchange software which would facilitate the conversion to and from the CCSM format through interfacing subroutines.

This software was then implemented at various federal and provincial agencies in an attempt to promote the Standards and to demonstrate their capabilities. The status of this effort is shown in table 1 (Piche, 1988).

The Alberta Bureau of Surveying and Mapping (ABSM) conducted a thorough evaluation of the file format to test the conversion process to and from CCSM of their digital map data at the scales of 1:1000, 1:5000, 1:20,000, 1:250,000 and 1:1,000,000. The test proved that this was possible but pointed to the need for the coding of textual information and parametric arcs. The recommendations made by ABSM and other organizations were implemented and version 1.2 of the CCSM File Format was produced.

In addition, the Topographic Mapping Division, EMR, set a policy for fiscal year 1990/91 whereby all data disseminated or acquired for inclusion in the National Topographic Data Base (NTDB) have to adhere to the CCSM Standards.

TABLE 1. Installation of the CCSM Exchange Software

Agency	Implementation date	System in use at that time	Code in use	Transfer format	Interest used
EMR Topo	04-87	VAX Intergraph	EMR Code	ISIF	High
Alberta (ABSM)	07-87	Vax 11/785 Intergraph	CCSM 82 version modified	DMDF	High
BC(MOEP)	07-87	VAX750 Intergraph	CCSM 84 version modified	MOEP	Low,wait for CCSM approval
LRIS	07-87	MICRO-VAX II CARIS	LRIS	DLG	High
CCRS	07-87	VAX Intergraph	EMR code		Medium
MCE	08-87	VAX Intergraph	MCE code		Low

Status of the CCSM National Standards
The CCSM National Standards serve as the basis for the exchange of digital topographic data between federal, provincial and private surveying and mapping agencies in Canada. The introduction of these standards will reduce costs through pooling the topographic information collected by the Canadian mapping organizations into a national data base. It will also facilitate the wide use of digital topographic data by providing capabilities that make it easier to use data bases developed by other organizations.

The data classification and coding standards and the quality evaluation have been adopted and used in Canada for more than five years. Software to convert digital data between the CCSM File Format and other formats were developed and implemented in several organizations, and the format for data exchange was evaluated.The results have been successful.

For the following reports:

National Standards for the Exchange of Digital Topographic Data:

Volume I: Data Classification, Quality Evaluation and EDP File Format:

Volume II: Topographic Codes and Dictionary of Topographic Features:

EDP File Format, Version 1.2:

Standards for a Digital Standards Topographic Information Model.

Contact Dr. M Allam, A/Director GIS Division, Surveys, Mapping & Remote Sensing Sector, Energy, Mines and Resources, 615 Booth St., Ottawa, Ont. Canada, K1A0E4:

MAP DATA INTERCHANGE FORMAT (MDIF)

The MDIF Project

The Ontario Ministry of Natural Resources has been developing the Mapping Data Interchange Format (MDIF) since October 1984. Over the years the specification of the format has evolved considerably. A draft of version 3.1 is being prepared (May, 1990) and will probably be submitted later this year (1990) to the Canadian General Standards Board as consideration for the basis of a national interchange standard. In earlier stages of its development the MDIF specification was closely linked to the Map and Chart Interchange Format (MACDIF) effort described in the next section.

MDIF Goals

The goals of the MDIF specification are (1) to provide an explicit data model that is flexible enough to handle current mapping needs, including those of remote sensing, civil engineering and geology, and to be extended easily to cover future needs; (2) to specify a formal syntax; and (3) to provide an efficient encoding for that syntax. The second goal overlaps that of the first by providing for a formalized completion of the data model through a careful selection of identifiers and the use of comments. It overlaps the third in providing the basis from which an efficient encoding can be made. MDIF is not intended for the transmission of maps (among which navigational, etc., charts are counted as special cases) as such, but rather for the transmission of the data that form the substratum of maps and of other geographically-related applications.

MDIF Description

The object being communicated in the MDIF data model is a *digital mapping package*, through which the data to be transmitted are organized into a *package header*, a set of *mapping data components* and a set of *associated information elements*. A package header contains administrative and coordinate-transformational information and definitions of attributes and of data items included in those attributes (a kind of data dictionary). Each mapping data component has its own *component header* which can complement or partly overide information in the package header and which is followed by a set of *feature descriptors*. The latter describes features that are point-like, line-like, area-like, and volume-like by means of attributes as defined in the package or respective component header and identifies them by means of one or more user-specified systems. (It is also possible to form descriptors of possibly non-homogeneous features or clumps.) It is not required that the delineation of features conform to the planar graph model. The only topological relationships used are those between an arc and its end-points and those between an areal feature and its outer and inner bounding arcs. Such a relationship is asserted in the definitions of the delineation of an arc or of an areal feature, and thus, is "married' to the geometry of a delineation and not separated from it. Coordinate systems used in the model are that of the reference spheroid (geographics), that of the mapping system (normally, rectangular cartesians derived from the projection plane and possibly its elevational axis), and that of an ad-hoc construct defined for each mapping data component (normally, rectangular cartesians, but always related to those of the mapping system by means of an affine transformation. The coordinates are constrained to be in the range [0.0, 1.0]). Associated information elements are similar to attributes except that they do not naturally pertain to a unique feature descriptor.

MDIF is specified in conformance with a number of existing international telecommunication standards, especially, the seven-layer Open Systems Interconnection Reference Model and the Abstract Syntax Notation One (ASN.1) specification which has been used in specifying the syntax for E-mail communications, for example. Currently, aside from the data model, the MDIF specification addresses communication issues pertaining only to layer six of the OSI Reference Model, i.e., the presentation layer, which is concerned with formatting of data. Services and protocols for the seventh layer, i.e., the application layer which for the purposes of MDIF consist largely of transactions with one or more geographical information systems, have been left for further study.

An outstanding characteristic of mapping (geomatic, geographically-related) data is the complexity of its data structures. This complexity defeats attempts to use a "flat file" format, i.e., one that provides for the encoding of primitive data items but cannot without some contortions handle data structures. ASN.1 provides two great benefits to MDIF: (1) its formal syntax allows for a formalized completion of the data model and (2) its encoding rules, which are logically separate from its syntax specification rules, allows for an efficient encoding of primitive data as well as for a representation of the data structures.

The following report is available from C. Broughton, Surveys, Mapping & Remote Sensing Branch, Ontario Ministry of Natural Resources, 90 Sheppard Ave. East, North York, Ontario, Canada, M2N 3A1:

Technical Specification of the Mapping Data Interchange

Format-MDIF.

MAP AND CHART DATA INTERCHANGE FORMAT (MACDIF)

The MACDIF Project
The concept of using telecommunication based standards to solve the exchange problems of mappers was initially proposed by the Surveys and Mapping Branch of the Ontario Ministry of Natural Resources. In June 1985, the Canadian Hydrographic Service (CHS) requested that the concept be extended to cover the exchange of navigational chart data and in particular to handle anticipated electronic chart applications. This led to cooperative work between the following agencies:

- Department of Fisheries and Oceans
- Department of Energy, Mines and Resources
- Department of Communications
- Department of National Defense
- Department of Supply and Services
- National Archives
- Ontario Ministry of Natural Resources
- U.S. Department of Commerce (National Ocean Service)

In late 1985 a project was initiated to take a different approach to the encoding of spatial data. It produced "MACDIF" for Map and Chart Data Interchange Format Version 2.0 (CHS, 1988) which was developed jointly by the Ontario Ministry of Natural Resources and the Federal Government. It was the first such geographic exchange format to be built on the "Open Systems Interconnection (OSI)" concept of the International Standards Organization (ISO). In 1988, on the completion of the project OMNR continued to work independently and their work is described under MDIF in the previous section. The work of the Federal Government has also continued separately and MACDIF has been revised to Version 2.3 (CHS, 1990) to include an up-dating mechanism which is of major importance to applications such as electronic charting and automobile navigation systems where up-to-date information is essential.

MACDIF Specification
MACDIF is compatible with the emerging international telecommunications standards for public data networks. MACDIF is a flexible format which can be used to communicate anything from raw digitized map information to a fully symbolized and cartographically enhanced map or chart. Annotation may be in English, French (with proper accents and diacritical marks) or any other language. MACDIF organizes information into a number of categories which define the overall structure of a

map, its relation to a world coordinate system, the features which make up the map, their attributes and boundaries and, optionally, any related symbolization and topological relationships. This data format allows for a blind interchange; i.e., there is only one single flexible format for encoding data which may be interpreted by various levels of receiving computer or terminal devices.

MACDIF can be viewed in different ways dependent upon the context in which it is discussed. From the technical point of view, it is a coding scheme for the representation and communication of map and chart data; i.e., it is a set of rules, a grammar, by which one may represent (encode) a digital description of a map or chart. The information is structured according to a rigourous, unambiguous syntax. This establishes a norm which forms the basis upon which to build a number of different independent applications, all sharing common data. This development of a general underlying coding scheme promotes the compatible development of a number of broadly-based applications. This is the essential element in the establishment and evolution of electronic-based mapping and charting applications.

MACDIF is also termed a proposed standard since it is intended that it be used as the common basis for a number of applications. The term "standard" is often misused to represent the specification of a commonly used coding scheme or other specification upon which systems or applications are based. However, the term "standard" more rigourously applies to a norm which has been established in an open public forum so that it represents the consensus derived from the consolidated experience of the industry. The development of a standard is carried out under the auspices of a national or international standards-making body according to certain well-established formal procedures and adherence to the principles defined by other national and international standards. This also provides stability and a mechanism by which a standard may be publicly maintained and updated.

The MACDIF specification is not itself a standard, but rather an input to the formal standards organization methodologies and procedures in order to assist in the establishment of universal, stable, public domain standards.

A distinction must be made between a coding scheme and any implementation which utilizes the coding scheme. There will be many other components required for an operational system in addition to the coding scheme utilized for the communication and possible storage of data. In many applications, such as in electronic charting, MACDIF will be an essential part but only one part of the application.

The MACDIF specification document (CHS, 1990) presents a description of a data interchange format for cartographic and hydrographic use which meets the needs of a wide variety of applications. The document presents the principles upon which MACDIF is based and then describes the specific structures which make up the interchange format. First, the overall information structure is described and then, the manner in which data is coded, is presented. Related developments which may affect the development of MACDIF are also referenced. These include efforts done in different contexts to produce survey and mapping or hydrographic interchange formats, as well as the on-going efforts to define data communications standards. A major focus is placed on handling updates.

The syntax of MACDIF is described in terms of the Abstract Syntax Notation (ASN) defined by the International Standards Organization (ISO) and is coded in terms of supporting data syntaxes defined in other international telecommunications standards. One of the principal features of MACDIF is the capability to code data in an efficient manner so that it may be easily telecommunicated. Each item of data, coded according to MACDIF, is identified by a TAG and LENGTH marker. Any portion of a map or chart can be easily communicated with little overhead. The same ASN encoded syntax applies to an update of a map or chart as to the whole map or chart. The difference is, that in an update, most of the information fields are optional and are included only if the particular information item must be updated.

MACDIF is designed to be a general standard for communicating map and chart data. It is intended both for professional use of mapping and charting agencies as well as the dissemination of information to industry and the public in electronic form. In order to achieve the maximum flexibility, it was important that the coding scheme be designed according to certain universal principles. These principles include:

- o independence from the hardware constraints of the equipment used within the applications,
- o independence from the media used for communications or storage,
- o communications transmission efficiency,
- o the ability to build upon other norms already established in the industry,
- o blind interchange,
- o meaningful defaults,
- o the ability to accommodate a wide variety of applications,
- o the capability to support multiple languages (including non-Roman scripts such as Cyrillic, Arabic

or Chinese),

- the ability to accommodate modification and extension in a forward and, where possible, backward compatible manner, and

- stability by being a public domain standard.

By communicating MACDIF positional information as fractions of a normalized unit coordinate system, a device-independent rendering may be achieved for use in a database or for presentation on the display screen of any real display device or on any plotter, i.e., the coordinate system used within MACDIF is based on a unit square (with a 0 to 1 range for X and Y). Parameters for a transformation are communicated, along with the data relating it to the real world coordinate system. Any presentation process may make use of the transformation specification and scale this data into its own device-dependent coordinate system. This approach to coordinate specification is also the most efficient manner of storing and communicating such data, since the number range matches exactly the area of interest and no extra digits are required to handle fixed biases. Coordinate and other information are packed into a small number of bytes while retaining the capability to specify these values to various levels of precision.

The MACDIF concept encompasses many different possible applications, including the interchange of basic source map or chart information from a producer to a central agency, the interchange of processed information between agencies, and the interchange of information to the public or industry in a form suitable for inclusion in a local data base or for presentation. Although data for all of these applications may be accommodated within the interchange format, different applications may use different aspects of the format. The handling of updates also differs between these various applications.

The prime concern in handling updates is the mechanism to ensure that updates are correctly applied to a particular map or chart. If there is a possibility that some updates may be missed, the integrity of the map or chart must not be corrupted.

In the MACDIF specification, three techniques are presented for the handling of updates. The first technique is based on organizing a map into cells and updating the map by replacing cells. The second technique is based on the communication of individual updates on an element by element basis. The third technique is a composite of the two others. The MACDIF specification can handle any of these three approaches. However, in an actual application, only one of the approaches would be used.

The structure of the MACDIF syntax consists of a

hierarchical tree of information elements. Each element is tagged so that it may be identified in a data set. The process which interprets (or parses) the data set, matches each data element in the data set to the type of data element expected by the syntax. Since each type of information element is identified by a unique tag, it is not necessary to include filler information in the data set for optional information which has not been specified.

Updating
An update message is structured in the same manner as a full MACDIF map or chart. The difference is that only a small portion of the information is included in a update message and that an update message must be combined with a base map or chart to have meaning. An update message is identified by the update flag being set. When the update flag is set, the MACDIF data fields in the Content-ID and Reference-ID sections (which are normally mandatory) are assumed to be the same as for the base map. The Producer-ID data fields (which are also normally mandatory) are optional for an update. The Data-Set-ID field identifies the update and the other fields contain the data pertinent to the update.

The general syntactic structure of an update is the same as that for a full MACDIF map or chart and consists of an Administrative Header which identifies the Data Set followed by Map Definition sections which contain data describing the components of a map or chart to be updated. An update must make use of the same feature, segment and node numbers as the base map to which it applies.

Status of MACDIF
Various agencies and levels of government cooperated to share information to develop MACDIF to meet their needs and the needs of industry and end users. The results of this research have been made available to the Canadian General Standards Board Committee On Geomatic Standards to contribute to the development of Canadian standards. They have also been made available internationally to various bodies, such as International Hydrographic Organization, for the development of international standards. The Canadian Hydrographic Service is also using it for the experimental exchange of Electronic Navigational Charts (ENC's).

The following report is available from: T. Evangelatos, Canadian Hydrographic Service, Fisheries and Oceans, 615 Booth St., Ottawa, Ont., Canada, K1A 0E6:

Specification of the Map And Chart Data Interchange Format: MACDIF, Version 2.3.

THE CANADIAN GENERAL STANDARDS BOARD (CGSB) COMMITTEE ON GEOMATICS (COG)

Background
Canada has a national standards system rather than a single

national standards organization found in most countries. At the core of this standards system is the Standards Council of Canada. The Council is a Crown Corporation and consists of 47 Canadians. It accredits and coordinates the efforts of 5 standards-writing organizations, of which the CGSB is one. It is also responsible for accrediting laboratories in various specialized fields, and in participating in international standards activities on behalf of Canada.

The COG Consensus Committee

The Committee On Geomatics for the development of national standards held its first meeting in June 1989 (Evangelatos, 1990). More than 43 representatives from industry, academia and government attended and some 40 agencies have followed up to become voting members. Another 35 persons currently serve as alternate or information members of the committee. Mr. Rene Gareau of the Canada Centre for Geomatics of EMR, was elected chairman of the Consensus Committee. The overall objective adopted by the committee was to "develop standards to promote the sharing of geomatics data". Four working groups were established in 1989 with a fifth group proposed in 1990. The overall priority for standards development was as follows:

- o Data Transfer/Interchange Formats (Working Group 1)
- o Data Models for the Transfer Format (Working Group 2)
- o Classification of Features (Working Group 3)
- o Cataloguing of Data Sets (Working Group 4)
- o Data Quality (Proposed W.G. 5)
- o Terminology
- o Geographic Referencing
- o Symbology

Working Groups

At the second meeting of the Consensus Committee, in October 1989, each of the four working groups presented detailed work plans and interrelationships among the groups were clarified. It became clear that the work of groups 1 and 2 must be closely coordinated, and that the feature classification group will also use the data model developed by group 2. Each group has begun by defining their scope, plan and criteria for establishing national standards.

Working Group 1 developed a set of requirements for a format, then evaluated existing formats against these requirements and found that no existing format fully met all the requirements. They are also looking at the possibility of a short term solution that would be functionally compatible with a more general, longer term, solution. They will use the model and approach adopted by Working Group 2 and are working closely with that group. The Geographic Document Architecture described at the end of this chapter has been adopted in principle.

Working Group 2 is establishing the evaluation criteria and has been examining existing models to see if they meet these criteria.

Working Group 3 is developing a broad classification schema which will be correlated with existing classification schema developed under the auspices of the CCSM. This work will be coordinated with that of working groups 1 and 2.

Working Group 4 is beginning with the development of draft standards for descriptive cataloguing of geomatics datasets. Since this work can be based directly on extensions to existing cataloguing standards for maps and machine-readable records, it is more straightforward than some of the other standards. A first draft working standard is anticipated by late 1990. However, efforts to develop electronic data dictionary/directory standards have been delayed through a lack of available personnel.

A fifth group has been proposed to deal with the difficult but extremely important issue of data quality. This group plans to use the American Spatial Data Transfer Specification as a starting point.

Leaders of the working groups are currently

- o Data Transfer: R. Baser, Department of Communications, Ottawa/R. Balser, Ministry of Crown Lands, Victoria.
- o Data Modelling: M. Sondheim, Ministry of Crown Lands, Victoria/C. Wong, Energy Mines and Resources.
- o Feature Classification: P. Friesen, Ministry of Crown Lands, Victoria/ M. Weiss, Surveys and Mapping, Alberta.
- o Cataloguing: D. Brown, National Archives of Canada;/J. Yan, Statistics Canada, Ottawa.

Several years of effort are likely required before the national standards are fully developed, tested and adopted. Never before have such broad-based standards in geomatics been undertaken in Canada. Wide support and participation, by members of the main committee, in the working groups will be required to ensure progress and then place the committee in a position to achieve consensus.

For more information contact J. Hillman, Standards Program Manager, Canadian General Standards Board, Ottawa, Canada, K1A 1G6

THE GEOGRAPHIC DOCUMENT ARCHITECTURE

Open Systems Interconnection (OSI) and the Office Document Architechture (ODA):
The process of interchanging information between two communicating entities has been analysed by the International

Standards Organization (ISO) Technical Committee 97, together with the UN based International Telegraphic Union (ITU) and International Telephone and Telegraph Consultative Consultative Committee (CCITT). These broadly based international committees have developed an architectural model for a communications or interchange process which breaks down an interchange into its component parts. This architectural model is known as the Open Systems Interconnect (OSI) model and is divided into 7 distinct layers. The OSI model is the basic building block upon which all standardized modern telecommunications systems are being built. It has world wide acceptance.

The Office Document Architecture (ODA) is an application layer (layer 7) standard (ISO 8613) for structuring office documents for interchange. It organizes a document, such as a word processor data file, into a number of layout objects consisting of the text content, any pictures, raster scanned images or other content types. The ODA is essentially a syntax, described in the Abstract Syntax Notation, for the structuring of office documents for interchange. Other OSI telematic services, such as the X.400 messaging services (International Electronic Mail), are compatible with ODA, and ODA documents may be carried by these services.

A Geographic Document Architecture (GDA)
In a similar manner to the definition of an Office Document Architecture for the interchange between office systems, a Geographic Document Architecture could be defined for use in interchanging digital geographic data. By building this GDA upon the OSI system of standards, compatibility is automatically gained with standard telecommunications networks world wide. Different content types may be defined to address the various specialized data types required and the data relationships required in digital geographic data interchange.

The Geographic Document Architecture can be thought of as a "vessel" which carries a number of different "containers". Any information may be put into the various containers. The GDA labels each of the containers. The vessel is carried across a standard "channel" which allows the transport of the GDA encoded mapping information. This architectural concept is identical to that used in ODA encoded office documents, X.400 Electronic Mail, etc. The communication "channel" is common to all of these applications. Figure 1 illustrates the GDA approach.

Information carried by the GDA will be coded at the presentation level using the following standardized coding techniques.

Basic Text	ASCII (or ISO 646)
Accents	Supplementary Characters (ISO 6937)
Other Alphabets	Registered Code Tables (ISO 2375)
Pictorial Information	Picture Coding (ISO 9292)

The GDA is intended to be a general interchange format for all types of geographic information. It therefore does not define precisely the contents of any of the "containers". The coding techniques and the general structuring of the information is common so that the geographic information can be easily read. However if a particular data type, feature code etc. is used on the source system and not available on the receiving system, the problem of how to represent the information in the receiving system is not solved by the interchange format. The information is carried by the GDA

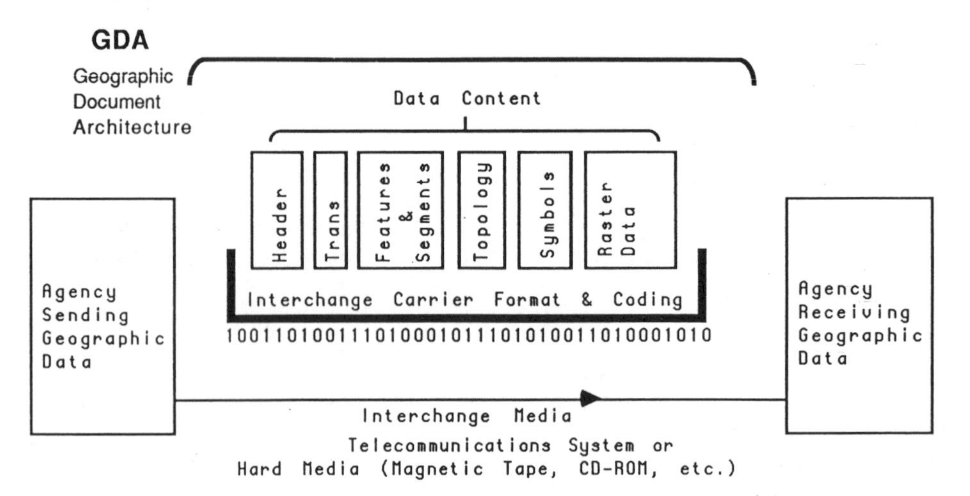

Figure 1: The GDA Interchange Structure

but there is no magic conversion which can remove an end-to-end incompatibility in the GIS application. In the interchange of information between GIS systems supporting incompatible functions, there are always end to end problems which cannot be solved by the interchange format. If the GIS systems are functionally compatible, then the complete interchange is possible through the GDA.

For particular applications such as the distribution of digital data to Hydrographic Electronic Charts, particular "Content Architectures" (in ODA terminology) or formats may be defined. Most importantly, content architectures can be defined corresponding to the information content of existing interchange standards. This means that the GDA will bring a large measure of compatibility, at least at the coding, general structure and communications levels, to all digital geographic interchange standards.

Telecommunications and Storage Media

The GDA facilitates the interchange of digital geographic data in a manner independent of the interchange media. The OSI model separates the information content at the application layer from the coding and delimiting structures at the presentation layer, and especially from the telecommunications media at the lower layers.

It is very common in OSI communication systems to substitute different physical, link and network protocols and standards. For example, digital data may be sent over an end-to-end data channel composed of wired links, microwave links and satellite links. The OSI defined lower layer standards obviously handle telecommunications systems very well. However, in the case of digital geographical information it is also important to be able to store and carry the data on hard media such as computer magnetic tapes, disks or CD-ROMS. It is, of course, possible to store the bit stream which would be communicated on a hard media. But other standards provide better methods of indexing into the data.

For telecommunications media, the OSI defined ISO 8824/5 (Abstract Syntax Notation) should be used to delimit and structure digital geographic information encoded according to the GDA. For storage media, the indexed approach specified in ISO 8211 should be used. It is entirely possible to convert between these two equivalent methods of delimiting and structuring the application records in the GDA.

Status of the Geographic Document Architecture

This effort is important for the development of compatible standards for the exchange of spatial data by both hard media and telecommunications. This work on the GDA concept is supported by the Directorate of Geographic Operations of National Defence. As research continues, results will be made available to appropriate standards bodies such as the Digital Geographic Information Working Group, the Working Groups of the Committee on Geomatics, etc.

For more information on the Geographic Document Architecture, contact: D. McKellar, Section Head, Geographic Requirements, National Defence Headquarters:, Ottawa, Canada, K1A 0K2.

REFERENCES

Allam, M.M., 1987, Canadian Council on Surveying and Mapping CCSM Standard, The Canadian Surveyor, Vol. 41, No. 3, Autumn 1987, pp. 443-446.

Allam, M.M., 1986, The Development of a Data Model and National Standards for the Exchange of Digital Topographic Data, Proceedings Auto Carto London, U.K., September, Vol. I, pp. 359-371.

Allam M.M., 1983, National Standards for the Exchange of Digital Topographic Data, Proceedings of the XVII Congress of the International Federation of Surveyors, Sofia, Bulgaria.

CCSM, 1982, National Standards for the Exchange of Digital Topographic Data, Draft Reports: I - Standards for the Classification of Topographic Features; II - Standards for the Quality Evaluation of Digital Topographic Data; III - EDP Standards Applied to Digital Topographic Data. Surveys and Mapping Branch, EMR, Ottawa, Canada,

CCSM, 1984a, National Standards for the Exchange of Digital Topographic Data, Second Draft Report, Volume I: Data Classification, Quality Evaluation and EDP File Format. Surveys and Mapping Branch, EMR, Ottawa, Canada.

CCSM, 1984b, National Standards for the Exchange of Digital Topographic Data, Volume II: Topographic Codes and Dictionary of Topographic Features. Surveys and Mapping Branch, EMR, Ottawa, Canada.

CCSM, 1985, Standards for a Digital Topographic Information Model, First Draft Report, November 1985. Surveys and Mapping Branch, EMR, Ottawa, Canada.

CCSM, 1986a, Standard EDP File Exchange for Digital Topographic Data, Version 0.0, First Draft Report / Revision to Second Draft Report Vol.I/ Part III. Surveys and Mapping Branch, EMR, Ottawa, Canada.

CCSM, 1986b, Proposed Standards for a Digital Topographic Information Model, Revision of First Draft, Surveys and Mapping Branch. EMR, Ottawa, Canada.

CCSM, 1989, National Standards for the Exchange of Digital Topographic Data, Revision of the Second Draft Report, Vol. I/Part III: EDP File Format, Version 1.2. Canada Centre for Geomatics, , EMR, Ottawa, Canada.

CHS, 1988, Specification of the Map And Chart Data Interchange Format: MACDIF, Version 2. Canadian Hydrographic Service, Department of Fisheries and Oceans, Ottawa, Canada

CHS, 1990, Specification of the Map And Chart Data Interchange Format: MACDIF, Version 2.3. Canadian Hydrographic Service, Department of Fisheries and Oceans, Ottawa, Canada.

Faucher, F., 1989, Status of the CCSM National Standards for the Exchange of Digital Topographic Data, Proceedings of the National Conference on Geographic Information Systems, Ottawa, Canada, pp. 381-390.

Evangelatos, T.V., Yan, J., and Haddon, B., 1990, Geomatic Standards in the Federal Government, Proceedings of the Second National Conference on Geographic Information Systems, Ottawa, Canada..

McKellar, D., O'Brien, D., and Lalonde, W., 1990, An Architecture for the Exchange of Geographic Data, Proceedings of the Second National Conference on Geographic Information Systems,, Ottawa, Canada.

OMNR, 1990, Technical Specification of the Mapping Data Interchange Format- MDIF, Version 3, Draft 3. Ministry of Natural Resources, North York, Ontario, Canada.

Piche, B., 1988, Overview of the CCSM Exchange Format, Unpublished Manuscript, Canada Centre for Geomatics. Surveys, Mapping and Remote Sensing Sector, EMR, Canada.

THE EFFORTS OF GIS STANDARDIZATION IN CHINA

Du Daosheng and Du Qingyun
Department of Cartography
Wuhan Technical University
of Surveying and Mapping
430070, P. R. China

INTRODUCTION

In the last decade, Geographic Information System (GIS) has been greatly developed in China, that has brought revolution in conventional cartographic technology.

At present, GIS technology is being widely applied to many fields relative to geo-science. Many research organizations is developing their own GIS. The development falls into two forms: One is to develop fully new information system software of themselves, another is to introduce the existent software such as ARC/INFO to establish regional GIS entity.

The development of GIS standards is a recent matter in China. It is the result of the deep and far-reaching research and application of GIS technology. The Laboratory of Resource and Environment Information System (LREIS) of Chinese Academy of Sciences, The Institute of Surveying and Mapping, The Institute of Standardization of National Bureau of Surveying and Mapping, and Wuhan Technical University of Surveying and Mapping (WTUSM) are some of the organizations that begin to get in touch with the problem. Their efforts can represent the overall appearance of the standardization efforts in China.

In general, the efforts are in the initial stage. Many of them are implied in broader project research, instead of constitute an individual project. This is partly because the cooperation among the organizations is not fully developed and that the importance of standardization hasn't been recognized. However, as can be seen from the following discussion, the problem is getting more and more attention in China.

SCOPE AND GENERAL GOAL OF THE EFFORTS

Cartographic Database and GIS

Ten years ago, when CAC technology just started its first step, the concept of Cartographic Database was often mentioned among researchers. This type of database has many functions such as input, storage, retrieval, edit and output of cartographic data. In a database, besides the topographic data are stored, some of other geographic data from thematic maps are deposited. It also has well-constructed internal structure and can store spatial relation of objects so that relevant analysis of topographic data with thematic data becomes possible. Gradually the boundary between Cartographic Database and GIS is no longer so definite. Many GIS softwares begin with a well-designed cartographic data-base and the key part in GIS data entity is cartographic data from topographic and thematic maps.

Scope and General Goal

As mentioned above, the major handling objective of both database and GIS is cartographic data, so most of efforts are concentrated upon the input, exchange and output of cartographic data.

Standardization of Data Input:

1. Sorting and Classifying of Geographic Features and Their Encoding

   We are fist confronted with the problem of how to sort and classify geographic data according to certain rules and to reflect them in GIS. For topographic map, the category and classification of features are standardized, so what should be done is merely to adjust them and give them correspondening codes. For thematic maps, the problem appears more complicated. The rules of sorting and classifying of the features are traditionally different because of the purpose, scale and area of the map, and their standardization is far from a simple cartographic problem, but needs the cooperation of experts in related specialties. However, a set of relatively stable standards of the sorting and classifying for the features is possible.

2. The standardization of some common geographic data

   The standardization of some common geographic data, such as national boundaries, polygon division within a country, including administrative division, geographic division, drainage division and so on, is very important to the information exchange between systems. If they are not standardized, even if no trouble results in specialty field, the share and integrated analysis of data from different systems is very difficult.

3. The standardization of data capture process

   Large scale data capture is with the help of electronic devices. If the data in field surveying are to be automatically handled, a set of unified data capture standards are needed.

   Digitization is an important way to capture data on map. The work contains more standardization problems. A specification for digitization is necessary, otherwise, when used the data results in much trouble of computation, could even become useless. Because of this technical limit, a compromise must be made between inputing information manually and deriving information by automatic computation for some input data. The former needs much more manual work of digitization, the latter more reliable software and computer time.

The Standardization of System Exchange: The work is to guarantee the flexible and convenient exchange of data among different systems. It has two prerequisites which may be regarded as two standardization problems. The first is that the data in the systems must be comparable, i.e., the data in one system is essentially useful and sharable to another system. For that most essential common data set, such as the number of geographic division with which the statistics is carried out, must be standardized. The second is that a unified exchange format must be available. This problem is relatively easy to solve. The importance is that the standard must reflect the property of each system on which the exchange proceeds.

The Standardization of Data Output: Two points must be considered here: one is standardization of map contents; another is that of map form.

1. The Standardization of Contents

   This is considerably complicated work, for it is concerned in the standardization of the map itself. The problem is generally easier for a topographic map because it has accepted contents. However that is not the case of thematic map. Under the concept of thematic map too much content is included. The contents of a thematic map varies with many factors. If the problem of standardization is specified with no regard of user's demand, the system will be out of touch with users and applications.

   It is not impossible to propose several sets of standards for one kind of thematic map after having considered the influencing factors thoroughly. The partial solution of the problem to some extent will speed up the making of a thematic map a great deal.

2. The Standardization of Form

Comparing with the standardization of contents, the standardization of form is easier. It includes the choice of representation method and the symbolization of the geographic contents. For the output of a topographic map a set of standardized symbol system should be available. In fact it has been so.

For the output of thematic map, because the same contents can adopt many different representation methods, the same method can be realized by more than one kind of symbolization scheme. The careless standardization will degrade the quality of the map, so it is inevitable to compromise between the efficiency and quality of map making. Multi-scheme in the standardization is a way to lessen the degradation.

HISTORY AND BACKGROUND

The development of Computer Aided Cartography is scattered and lacked a concentrated plan in its initial stage, the thereby resulting in the demand for standardization. In China, most of the work of standardization is for the work division and cooperation among the partial organizations, and that what has been accomplished by all of them can serve the same purpose with no resulting communication obstacle. In most cases, the work is a part of the cooperation project. The standards are worked out by the cooperating organizations and observed by them. They are accepted gradually by other organizations later and become the commonly adopted standards.

Up to now, there is still no formal GIS standard promulgated in China. However some independent projects related to standardization are being carried out or have been finished. It is believed that the final results of the projects will become the national standard after further discussion and application testing.

DESCRIPTION OF THE CURRENT EFFORT

Conceptual Background

In China, little work is done on the development of the concept system and theoretical frame about GIS. Essentially the model of an object is taken as the basic processing unit in GIS is inherited. From the view of GIS software, the object is the basic logical unit that GIS can handle, to which all geographic information is adherent and all handling operations are applied.

From the view of geographic information, the object is the representation of a geographic entity. It is defined for a relatively independent concept on a map. In fact the definition of an object is very dependent on the demand of information processing.

An object includes feature code, geometric coordinate and name, and some quantity, quality and relation information. Feature code stands for the category and classification of the object, and is subordinated to the standardized code system of category and classification. Geometric coordinate string reflects the location of the object. Name is the text of placename about the object on the map. Quantity and quality information is some labeling information about object such as height of bridge, width of river, etc. Relation information reflects the geometric and topological relation between one object and another object.

Project efforts

As mentioned above, there are some independent projects related to standardization that are being carried out or have been finished. The progress in these projects represents the current situation in GIS standardization development in China.

The Code System of Category and Classification of Land Basic Information: The project is carried out by The Institute of Surveying and Mapping and The Institute of Standardization of National Bureau of Surveying and Mapping. During the research, many experts are invited to participate in the making of the standard.

The main purpose of the project is to serve the project Land Basic Information System going ahead in the institute. Now the standard has been worked out. The standard land basic information (mainly topographic map features) is sorted into nine categories and each category has several classifications. Then the encoding is carried out upon the categories and classifications.

The Standardization of Resource and Environment Information: While the Branch of Surveying and Mapping is concerned about the standardization of topographic map features, The Laboratory of Resource and Environment Information System of Chinese Academy of Sciences is dealing with the standardization of so-called resource and environment information, which mainly consists of widely distributed geographic thematic data. This project is undertaken by LREIS. It has immediate and ultimate goals. Its immediate goal is the standardization of resource and environment information, and ultimate goal is that of information system software itself.

The standardization of information covers the whole scope in the previous section. For example, many research sections will be confronted with the division of drainage area in China during the establishment of the system. For the comparability and exchange of data between different parts of a system, the division must be standardized. The standardization of the system tries to unify the key part of the system, such as data model, data structure, user interface and so on. At present the project is in progress in coordination with other application projects in the laboratory.

The Systems Engineering Design of Thematic Map: This project is undertaken by the Department of Cartography, WTUSM, sponsored by National Bureau of Surveying and Mapping.

This is a project in that standard and system development is equally important. Its first goal is to standardize the contents and form of common thematic map, i.e., for each type of thematic map, one or more sets of scheme about the contents, correspondent representation and symbolization should be formulated. Then a software system for the making of thematic maps should be designed based on the standard. In the past the batch computer aided making of thematic maps was a difficult task because the thematic map is so irregular that no standardized and universal software system is available. The project will try to solve this problem.

The work of the project will be assisted by The Institute of Standardization, for the standardization of thematic maps is also what they are concerned with. Their participation will benefit the authority of the standardization. Now the research group is just formed and begin the research.

The Creation of 1:4,000,000 Geographic Base Database: The project is cooperated by The Department of Cartography, WTUSM and The Map publishing House of China. The base data includes national boundaries, main rivers, cities and other features. The created database will be the standard of the geographic base of all the maps to be published in China. It can be expected that the database will become the information standard of the related GIS. Now the research group of the project have worked out the specification of digitization and tested it with smaller scaled maps. The digitization will be carried out later.

Computerization of Topographic Map Symbols. This is a project that has been finished by The Institute of Standardization of National Bureau of Surveying and Mapping.

In the conventional topographic map production of China a set of standard symbols is available. Because the standard is for manual map drawing, the symbol system appears too complex for computer aided mapping. Based on that, the researchers adjusted and improved some symbols in the original standard to make them simpler and regular thus more easy to draft by computer. At the same time, a topographic symbol base has been built up. The new symbol standard is only available for larger scale maps, and is expected to become the tool for the standardized output of large scale maps.

The Creation of Regional Topographic Database. This is a project undertaken by the Department of Cartography, WTUSM, sponsored by National Bureau of Surveying and Mapping. The system software will be developed based on the previous work of the Department. Because the software will be used by the different sections that undertake the task of digitization, the standard exchange interface is very important. Now a standard exchange format is being used. Digitization is a vast engineering project for the creation of the topographic database. Under this situation the work is divided and undertaken by different sections. Hence, the digitization specification is especially important. The standardized and formalized digitization rules is necessary. As can be said, in this project, the system is standard. The input, storage, retrieval, edit and output is realized by the same system. Then the standardization of information should be guaranteed by the digitization specification, such as how to define an object, whether a location point of place name should be input, and so on. The project is in progress at present.

The projects described above are some that represent the current development effort in China. Some of them are theoretical standardization problems, some are subordinate standardization problems serving a broader project. As can be seen, GIS standardization is getting more and more accentuated in China. It is believed that a bright prospect is promised.

CONCLUSION

Although we have seen optimistically the continuous development of GIS standardization, to achieve the ultimate goal there is still a long way to go. This is because not only the effort of standardization must be based on the experience of the design and establishment of GIS (but this will take a long time), but also a procedure of improvement and popularization is indispensible even after the standards have been worked out.

The effort of GIS standardization is initial. The research is not very far-reaching. Now the authors can find no published materials about the subject. But the authors believe that the organizations mentioned in previous sections will be pleased to provide us with more detailed further information.

THE STANDARDIZATION OF GEOGRAPHICAL DATA INTERCHANGE IN FINLAND

Antti Rainio
Project manager
LIS-project/National Land Survey
Helsinki, Finland

ABSTRACT

The development work of standards for geographical data interchange was started in 1985. It has produced national administrative standards based on Edifact (ISO 9735), but also products supporting the transfer mechanism. The geographical data dictionary system as a server and the conversion software package will secure the standard interchange.

INTRODUCTION

The current technological push is changing not only the work of map suppliers but also the work of map users. LIS/GISs have been implemented in several applications during the last few years.

In Finland, computer assisted cartography was started in the late sixties by producing thematic maps concerning census data. Computer aided topographic map production was started in the mid seventies.

In the eighties, the map production systems turned to GISs. Large programs of collecting geographical information have been launced. Decisions have been made to add coordinate data to the earlier register systems in order to share the benefits of coordinate linkages.

This development has occured in many sectors and organizations at the same time. Therefore the need of standards for geographical data interchange has arised strongly. Even if we still live the era of data collection and found the basis, it is important to make plans and test methods for cooperation in the future.

In Finland, the standardization project, so-called LIS-project, was launced in 1985. From the very beginning, the work was based on decentralized maintenance and the joint use of geographical information.

SCOPE AND GENERAL GOALS

Overall Aim

The idea of standardization is to separate information and systems from each other. The information as a resource is chancing slowly. The generations of systems are much shorter than the generations of information.

The standardization of data interchange is a task to standardize the whole process - not only the interchange format. The data interchange process is always duplex. The user has to express his/her data needs and the supplier will send the asked data.

The standardization is a good start for cooperation and rationalization. In the future, GISs will not be separate but they will more and more interact. The idea of the joint use means that organizations and their information systems must develop as a part of the whole society.

Geographical Information

Geographical information is a unity of spatial and attribute information describing an entity or a phenomen of the real world. It is neither pictorial information nor combound to a scale as the cartographic information is.

- GEOGRAPHICAL INFORMATION
  - Spatial information
    - coordinates
    - geometry
    - topology
  - Attribute information
    - identifiers
    - linkages
    - time expressions
    - descriptive data

Geographical information is based on coordinates while land information may include only tabular data and be based on administrative identifiors. However, the entities described in GISs and LISs are basically the same, so the method to build up an object should be general enough.

Reality - Conceptual Model - Data Store

Everybody has his/her own conceptual model of the one and only reality. The tradition of information technology has claimed to avoid redundancy in data store and users have been forced to identify a common basis for object develop and market Esperanto.

In the information society the data store is or will be decentralized. The data are maintained by various authorities to serve mostly their own data needs and processes. From the users' point of view the conceptual model of all geographical information should be a one whole. So, one basis for practical modelling is a try to integrate the concepts of the existing databases and files.

REALITY	CONCEPTUAL MODEL	DATA STORE
Entity	Object type	Object
Attribute	Attribute type	Attribute value
Shape	Geometric object type	Geometric object

Cartographic representation has a long tradition. However, geographical information has a different character because of the claim for the integrity of the model and a great number of attributes connected to the geometric objects. To some extent the rules for conceptual modelling of static and dynamic environment or reality in general do lack.

Standardization Areas
In order to arrange the geographical data interchange the project was decided to make proposals to standardize the following topics (Rainio, 1986):

- o geometric object types
- o coordinate system and positioning accuracy classification-attribute classifications and coding systems
- o data representation and description
- o geographical query language

It was not only a question about the format but also about the principles of the joint use concept. To make the work thoroughly a wide inventory of data resources and administrative matters was to be drawn up.

HISTORY AND BACKGROUND

Earlier Efforts
In the late seventies, preparations of computer aided cadaster systems were going on. The need of various geographical information concerning parcels was obvious and the ideas of centralized land information system were painted. The possibilities of the coordinate linkage were realized but the time had not yet come to implement them.

In 1983, the "LIS-group" with representatives from

eleven organizations started to clear up and to plan the standardization work which had to be done. In 1984, the group made a proposal for the Ministry of Agriculture and Forestry to launch a LIS-project. The goal was to outline the joint use of geographical information on the basis of decentralized data maintenance and to prepare standards for data interchange.

At the same time, a so-called FINGIS-format was made up and brough into use. It is a simple and non-efficient fixed format for representing points and lines. In spite of its limitations in expressing attribute data it has been widely used in interchanging cartographic data.

LIS-project

In 1985, the Ministry of Agriculture and Forestry appointed a steering committee for the LIS-project (Rainio, 1986). The work of the project involved wide cooperation. Almost one hundred people from about 30 organizations have participated in eight working groups (w.g.). The practical work has been carried out by the National Board of Survey.

Efforts	RESULTS
pre-studies	
launch of the LIS-project (Steering Committee)	
geometry w.g.	VHS 1040
positioning w.g.	VHS 1041
classification w.g.	
administration w.g.	
data resources w.g.	GDDS
data contents w.g.	
data transfer w.g.	G-EDIS
query language w.g.	
1985	1990

GDDS	Geographical Data Dictionary System
G-EDIS	Geographical Electronic Data Interchange Software
VHS 1040	Standard for data description
VHS 1041	Standard for geographical data representation

The work of the groups was phased for three years. After that, a testing period was started in order to harmonize the proposals and to prepare the joint use practice and the national administrative standards (Ahonen, Rainio, 1989).

The principles of geometric representation were established quite soon. The chosen syntax was new and unknown, so a short while was needed to adopt and to be convinced of it.

After the first inventories the impossibility of an all covering classification was realized. The phenomenal world of geographical information was found wide and infinite contrary to the printed map. Therefore, the task was to organize the description of data resources.

Administrative problems such as data access, data secure, prizing policy and copyright were recorded.

The project has had several full time workers which has been necessary in order to secure a homogenous result and a good start for implementations. On the other hand, the standardization process was, to a great extent, a task of education.

General EDI and OSI Standardization Activities
EDI (Electronic Data Interchange) has mostly been developed for international commerce and trade purposes.

The Edifact (Electronic Data Interchange For Administration, Commerce and Trade) standard has been confirmed in ISO (ISO 9735), and it gives fairly open rules for any purpose data representation.

Edifact is a syntax for structure data represented by a certain character set. Interchange structure can be defined very fluently using repetition and nesting capabilities. The data representation is based on tags and separators, and also rules for omission and truncation are included, so the representation is robust and efficient.

In the next few years, X.400 and X.500 seem to be the basis for data transfer in networks. Within a few years, X.400 will also support the Edifact standard especially. Private sector and public administration in Finland have together outlined the FOSIP (Finnish OSI Profile).

Geographical data interchange should follow these general trends as much as possible.

RESULTS OF STANDARDIZATION

Overall Results
The LIS-project has produced national administrative standards (VHS) for geographical data representation and data description:

- VHS 1041 Geographical Data Representation
- VHS 1040 Message Description

The standards are based on Edifact syntax and they form a framework for data suppliers to define their data for the joint use. The standards give a general interface which is open for any data set or for any operative system. When the standards are supported by suppliers the data resources will together form a logically one database even if the data are maintained in various GISs.

The standards support conceptual modelling by:

- o defining geometric object types
- o providing a uniform model to describe various data sets.

Besides these standards, the project has produced a conversion software and a geographical data dictionary system.

Geographical Data Representation
The national administrative standard for geographical data representation (VHS 1041) covers:

- o reference to the coordinate system
- o positioning accuracy classification
- o geometric object types and their representation rules
- o attribute data representation rules
- o reference to the data description standard

The national coordinate system will always be used, and the format for coordinates has been defined. The positioning accuracy classification is recommended to be used with the coordinate representation (Vahala, 1986).

The geometric object types are:

- o point, chain, area
- o grid and grid cell

The geometric objects are defined to be represented by the Edifact data structures (data segments and data elements). An area can be represented as a simple/complex polygon or as area boundary chains with a reference point. Data suppliers are recommended to define area information of the both types if possible. When a grid and cells are used, a grid shall always be oriented by the north axis of the coordinate system. The definition allowes to represent topological identifiers for vector type objects.

The VHS 1041 defines the method for representing attribute information connected with spatial information using the Edifact data structures. The example of an area data definition sceme shows that the standard allows to define hierarchical object structure and attributes can be connected with geometric objects in all levels. The boxes in the sceme represents data segments with certain data elements which have been defined in VHS 1041 or in a certain data definition.

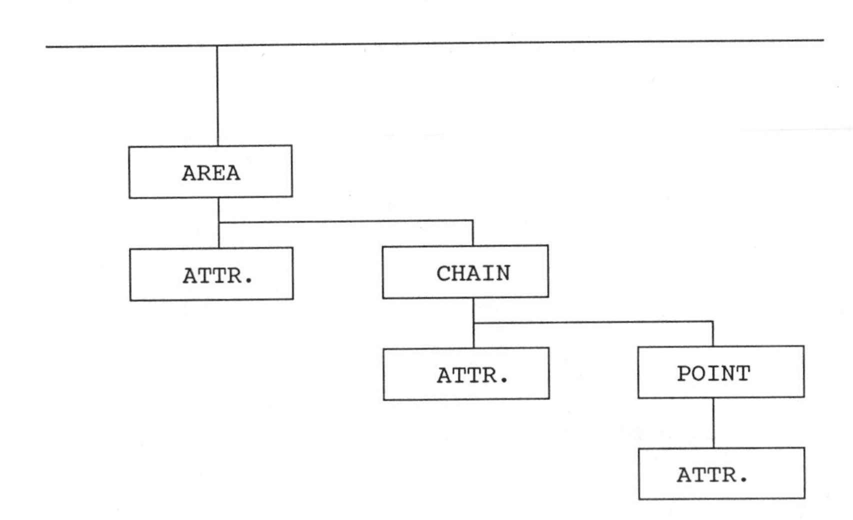

Data Description
The Edifact syntax itself does not include any method for data description but it is possible to define an Edifact-message for that purpose. National Administrative Standard (VHS 1040) has been developed in order to provide a standard mode for data description (Rainio, 1988). (It is confirmed to be used even in other applications than the geographical data interchange.)

The VHS 1040 is a general definition for describing any data which will be interchanged by using the Edifact. The description includes both the data structure and the data contents, i.e., the semantics such as the explanation of code values.

Geographical Query Language
As an interactive process the geographical data interchange needs a query language. The user will express his/her data needs in a standard mode and the supplier's system will automatically pick up the data, and convert and send it.

The query language (Ahonen, 1990) does base neither on any database management system nor on the SQL. The language is built on relational algebra, common geometric objects and geographical operations. The set of potential operations is very limited because of the lack of research in this field. Furthermore, in practice, the capability of GISs to support the operations varies.

A query concerns the entity types which are defined in the standard form and described in the standard way. To integrate the data representation and the query language the Edifact syntax has been examined as a basis for first implementations.

IMPLEMENTATION OF GEOGRAPHICAL DATA INTERCHANGE

Importance of Support

Standardization work does not succeed if the support of introduction is neglected. Efficient marketing is important but actually the cooperation is even more essential. It has to be secured that basic tools supporting the standard exist and are useable, and the users are aware what has to be done.

The public organizations can support the standard also by prizing the data lower when the standard is used in the interchange. The software houses will be very interested of standard when the byers require it to be supported.

Geographical Data Dictionary System

The geographical data dictionary system (GDDS) (Rainio, 1990) contains descriptions of the information available. The dictionary is a server telling the owner, entity types, attributes, spatial characters, and areal coverage of every data set. Furthermore, it provides tools for ordering the data.

The online dictionary system has both textual and graphic user interface with index maps. The dictionary provides tools for browsing the data descriptions and areal coverage.

The dictionary describes all entity types and their all attributes in the standard form. The data descriptions will also be published as a printed catalog.

Data Conversions

The conversion is required at the both ends of the inter-change. The LIS-project produced a conversion software G-EDIS (Geographical EDI Software) in order to get a wide experience in applying Edifact for geographical data (Rainio, 1988).

The software package is an Edifact interpreter but it has also some basic tabular and functional conversion capabilities. For example, coordinate transformations and changes in code values are possible.

The software reads the standard message description to control the conversion.

CONCLUSIONS

The standardization process takes a long time. That is why the results have to reach into the future. The results are still not allowed to be too advanced, otherwise they will not be understood and brought into use.

It is important to look around what is generally happening and to choose the best carrier waves which will quarantee the best support. Standards must be based on standards, although it might seem much easier to create everything by the scratch.

We are moving to the information society where the data are a material for many processes. To serve geographical information in compatible form it will be an advantage and it will have a wide influence on the development.

REFERENCES

Ahonen, P., Rainio, A., 1989, "Geographic Information in Joint Use - Cooperation Project in Finland", Proc. AM/FM Nordic conference I, Beito, Norway.

Ahonen, P., 1990, "Query Language for Geo-information - A Challence in Developing the Joint Use of Geographical Information", Proc. FIG XIX Congress, Helsinki, Finland.

International Organization for Standardization, 1988, "Electronic Data Interchange for Administration, Commerce and Transportation (EDIFACT) - Application level syntax rules", ISO 9735.

Ministry of Finance, 1990, "VHS 1040 Sanoman kuvaaminen", Valtion painatuskeskus, Helsinki, Finland. (National Administrative Standard 1040 Message Description; unofficial translation in English is available)

Ministry of Finance, 1990, "VHS 1041 Paikkatietojen esittäminen", Valtion painatuskeskus, Helsinki, Finland. (National Administrative Standard 1041 Spatial Data Representation; unofficial translation in English is available)

Rainio, A., 1986, "The LIS Project in Finland", Proc. FIG XVIII Congress, Toronto, Canada.

Rainio, A., 1988, "Spatial Data Exchange in Finland - The Edifact protocol and Data Conversion", Proc Second Scandinavian Reseach Conference on Geographical Information Systems, Honefoss, Norway.

Rainio, A., 1990, "The Geo-Data Dictionary System", Proc. FIG XIX Congress, Helsinki, Finland.

Vahala, M., 1986, "Data standardization of an integrated LIS in Finland", Proc. Auto Carto London, UK.

Further information: LIS-project/Antti Rainio
National Land Survey
P.O.Box 84
00521 HELSINKI
FINLAND

phone + 358 0 154 3581
fax + 358 0 147 289

LA NORMALISATION DES ECHANGES D'INFORMATIONS GEOGRAPHIQUES EN FRANCE

François Salgé
Ingénieur en Chef Géographe
vice-président de la commission
"normalisation des formats d échange" du CNIG
Institut Géographique National - France
BP 68
94160 Saint Mandé
France

ABSTRACT

The french activity concerning exchange format is conducted towards two principle direction. The first one is strategic: the CNIG working group have to be recognize as the only group in France where geographical information exchange systems are discussed, and the European context have to be taken into account. The second one is technical: the works deal with data model (theoretical concerns), content definition (semantic concerns) and exchange systems (syntactic concerns).

Each paragraph is presented by a short summary text in English and its development is written in French.

INTRODUCTION

Le Conseil National de l'Information Géographique
The National Council for Geographic Information (C.N.I.G.) is a governmental organization attached to the Ministry of Planning. This consultative organization is in charge to contribute to the development of geographic information uses in France by taking into account the needs of public and private users. It is mainly involved with the study of the capture and identification of spatially-referenced data, their analysis and their processing, and the definition, production, conservation, and distribution of all derived products.

Six commissions have been set up:

- o commission on large-scale topo-cadastral mapping
- o standing commission on geographic research
- o commission on evaluation of the economical and social utility of geographic information
- o national commission on toponymy
- o commission on the "national digital ground survey"
- o commission on exchange formats standardisation

La commission permanente de la recherche géographique
The standing Commission on geographic research is involved with the evaluation of what is going on in the research program of the French organisms dealing with geographic research, with the definition of general objectives of that research, and with a mutual information and coordination between partners.

An inquiry upon the geographic information community (2000 samples, 622 answers) was done in 1987. Its purpose was to get remarks on a draft definition of the "large scale topo-cadastral mapping". One of the main conclusions was the general need for the definition of a standard exchange format.

During the first symposium organized by this commission, held in Paris (13-04-88), all the 15 speakers were more or less claiming the importance of the definition of a suitable way to exchange geographic data.

This context led the CNIG to set up, in May 1988, an exchange format working group.

La commission "Normalisation des formats d'échange"
One of the working groups of the permanent Commission on geographic research was involved with the definition and the proposition of "an exchange format standard for digital geographic data".

The work was essentially to develop a system of exchanging digital geographic data which is able to give intercommunicability, interoperability and compatibility within the producers and users of geographic data in France. This includes the documentation of the whole standard.

It was set up in May 1988 and during 1988-summer an enquiry among professionals concerned with geographic information was undertaken to establish a list of the formats in use in France. Following discussions by that group it was decided to charge the group with defining a standard according to the following steps:

Definition of the final document
Digital Geographic information (DGI) logical structure
physical structure
annexes (coding system, glossary, etc..)
Definition of theDGI logical structure
Definition of the standard
Definition of a coding system
Study of the means of diffusion and approval of the standard.

This Working Group is composed of some 20 representatives of various organisations of that domain and is assisted by a group of 7 experts involved in the technical tasks.

It has been clearly identified to study the existing standards in the world in order to learn from other experiences, and to avoid the known pitfalls. The first lemma is that the data models for geographic information are scale-and-nation-independant. The second one is that the way to apply models are nation-independant even if the results are nation-and-scale dependant. The third one is that the exchange standard documentations have to be fully understood (French texts are mandatory). The last one is that there will be in the future a need for international standards. Those lemma imply to work within the knowledge of what is going on in an international basis.

Actually the elaboration of a French exchange standard cannot be separated from the one of an European exchange standard, and both actions have to be led at the same time.

This working group is now known as the commission "Normalisation des formats d'échange" and is seen as a separate body. (April 1990 decision of the CNIG plenary assembly).

OBJET ET OBJECTIF DE L'EFFORT NATIONAL

This paragraph deals with the purpose of the national work and sumarizes its outlines.

Pour l'essentiel, les travaux consistent à élaborer un système d'échange de données géographiques assurant un degré nécessaire d'intercommunication entre systèmes informatiques, et de compatibilité entre les différents centre producteurs et utilisateurs français de ces données. Simultanément, il produira un document de synthèse permettant au lecteur de comprendre et d'utiliser aisément le système en question.

Ces objectifs imposent cinq conditions:

1. Une telle norme doit échanger toutes structures de données possibles (DAO- spaghetti- topologique- maillé)

2. Les échanges physiques doivent pouvoir être faits par

bandes magnétiques disquettes, RNIS, etc...

3. La documentation DOIT etre lisible

4. la méthode de présentation des données doit être commune (mêmes concepts de base) tout en acceptant des variantes (souplesse)

5. dès l'en-tête sont intégrées des informations permettant d'interpréter automatiquement et sans équivoque la structure des données.

Le cahier des charge de l'élaboration de la norme comporte six contraintes:

1. rapidité de mise en uvre

2. économie de moyen impliquant en particulier l'analyse de normes existantes ou en cours de mise au point

3. modernité impliquant la prise en compte de modèles de données validés sur le plan théorique et permettant de suivre les évolutions du génie logiciel

4. dimension internationale et tout particulièrement Européenne

5. maîtrise du produit: participation active de la France à l'élaboration et l'évolution d'une norme Européenne

6. prise en compte des rôles respectifs du CEN, de l'ISO et de l'AFNOR.

DESCRIPTION DE L'ETAT D'AVANCEMENT

volet stratégique
This paragraph shows the strategic dimension of what must be done. This implies to develop the French organisation, and an European one. As each of them are specific, a short presentation of the context and the specific goals is made followed by the current status of the work.

Organisation française: The CNIG (National Council for Geographic Information) is the main advisor of what should be the technical standard. AFNOR is the French body for standardisation and thus the main advisor of what should be the best route to the effectiveness of the standard (in the context of EDI-France). The specific goals of what should be the role of the French geographic community leading to the adoption of geographic standards is then explained. As far as things are going on the CNIG commission on exchange format will be part of EDI-France organisation by the end of this year. The principle of such a partnership will be adopted by both organisms by the end of July.

Au sein du CNIG la commission a été organisée en un comité

de pilotage et une équipe de spécialistes.

Le comité de pilotage est composé de représentants officiels des ministères ou organismes concernés[1], il est chargé
de fixer les objectifs pratiques à atteindre,
de suivre l'avancement des travaux,
d'en évaluer les résultats,
d'assurer l'ensemble des relations avec les organismes concernés,
d'informer la communauté géographique française (et Européenne) de l'avancement des travaux par le biais des salons, colloques et revues,
de coordonner les actions de la France dans les instances internationales.

L'équipe de spécialistes de l'information géographique numérique, mis à disposition par certains des organismes[2] de l'équipe de pilotage est chargée de réaliser les tâches techniques définies par le comité de pilotage au cours de sessions.

Dans la mesure où la promulgation de normes est en France du ressort de l'AFNOR (Association Française de NORmalisation), un rapprochement avec celle-ci a ete des le debut pris en compte. Au sein de l'AFNOR, une entité spécifique disposant d'un budget propre a été créé en 1990 (EDI-France). Elle a pour but de favoriser l'Echange de Données Informatisés et donc de faire avancer les travaux de normalisation touchant à l'EDI au sens général.

La commission du CNIG est de-facto le lieu de rencontre de tous les utilisateurs d'informations géographiques numériques et doit devenir à terme le point de convergence de tous les travaux français, voire de certains travaux européens. Toute

---

[1]ministères: finances (cadastre), agriculture et forêts, défense, équipement, intérieur
organismes: IGN, CNIG, AFNOR, EUREKA-FRANCE, Service Hydrographique et Océanographique de la Marine, Bureau de la Recherche Géologique et Minière, Association des Ingénieurs des Villes de France, Centre National de la Recherche Scientifique, GIP-RECLUS, Electricité De France, URBA2000, FRANCE-TELECOM
privés: Ordre des Géomètres Experts, Chambre Nationale Syndicale des Topographes Photogrammètres, Chambre Nationale Syndicale des Photogrammétres Privés

[2]ministères: finances (cadastre), agriculture et forêts, défense, équipement
organismes: IGN, BRGM, CNRS, GIP-RECLUS
privés: OGE

prénorme d'échange d'informations géographiques numériques en provenance d'autres nations, tout avis sur ce type de prénorme et toute proposition française doit passer par la commission du CNIG.

Ainsi le but spécifique de l'organisation française est de faire reconnaître la commission sur les formats d'échanges du CNIG comme groupe opérationnel d'EDI-France afin de devenir ce point de passage obligé.

orientation Européenne: The European context is specific as there exists a European body (within EEC) which is involved with the definition of European standards. The exchange problem in Europe is now an European one as a lot of studies are cross-boundaries.The specific goal of this European environment is to promote the organisation of an European working group dealing with the set up of an European exchange standard. Some informal contacts are already set-up. The French delegation to next conferences will be able to present the French view of that potential European standard.

L'organisme européen officiel de normalisation est le Comité Européen de Normalisation (CEN). La particularité de ce dernier est que toute norme adoptée par lui devient obligatoire pour les pays de la CEE. Les organismes normalisateurs des pays doivent alors annuler ou modifier les normes nationales qui viendraient en contradiction. Ainsi au delà des normes adoptées par l'ISO, celles du CEN ont un enjeu stratégique important.

L'Europe de 92 en particulier et l'Europe au sens géographique sont deux réalités de la communauté géographique européenne. Quelques exemples viennent illustrer cet état de fait. En premier lieu les efforts dans le domaine de l'assistance à la conduite automobile (Prométheus dans le cadre EUREKA [Conseil de l'Europe] et DRIVE [CEE]) consistent entre autres à concevoir des bases de données routières par nature "Pan-européennes". En second lieu le projet CORINE de la CEE où la constitution d'un système d'information sur l'environnement pose le problème d'une base de données Européenne. On peut citer également les efforts du CERCO dans ce domaine.

Le groupe de travail , à rattacher vraisemblablement au CEN (?), est chargé de faire emerger une norme d'échange européenne à l'horizon 1995 qui soit acceptable (et acceptée) par tous les intervenants de l'information géographique numérique (producteurs, équipementiers, utilisateurs).

volet technique
From the technical point of view, exchanging data can be divided into 4 main problems.The fist one is "interfacing conceptual models" (data organisation). The second one is the cross referencing of content (data definitions). The third one is the definition of the exchange system for files. The

last one is related with the exchange of messages. Those two last topics can be seperated into two layers: the geographic layer and the transmission layer.

<u>modèle conceptuel des données:</u> The main assumption is that, conceptually, there exists a theoretical conceptual model able to modelize any geographic data set: how the data to be transmitted are assembled from the semantic point of view. The transmission model which explains how the data are transmitted is not related to that topic.The specific goal of this study is to put forward the universal conceptual model as it is presently within the general knowledge of the various actors of digital geographic information.This activity is already finished and further research will be done when new requirements will appear. This conceptuel data model will be compared with other conceptual models already defined (geographic data model for road information, ...) in order to produce models applied to technical domains.

Dans le contexte français l'état de l'art conduit à envisager l'échange de données maillées, CAO/DAO, spaghetti et topologique-structurée. Ainsi un modèle de données "universel", permettant de caler tout modèle de données spécifique, est à mettre au point. Pour un échange donné entre deux systèmes, la mise en correspondance respective de chacun des modèles de données avec le modèle de référence permet d'envisager un interfaçage avec le format d'échange qui soit indépendant des conditions d'échange.

L'étude faite a conduit à proposer un modèle de données général ("universel") qui puisse modéliser tout type de jeu de données géographiques. Ce modèle, mis au point en tenant compte des connaissances au niveau recherche et des besoins clairement exprimés, a été confronté aux principaux modèles existants: CAO/DAO, spaghetti, topologique.

Ce modèle théorique repose sur des objets élémentaires (arcs sommets faces) portant la géomètrie, formant un graphe planaire et permettant de construire des objets complexes. Des attributs géographiques peuvent être associés aux objets complexes et des liens sémantiques permettent d'exprimer des informations de type "relation entre objets".

Dans la mesure où la géographie recoupe de nombreux domaines techniques, pour chacun d'eux on peut définir une façon spécifique de modéliser les données en se basant sur le modèle "universel". Pour chaque domaine technique identifiés (réseau routier, ville, agriculture), des règles d'utilisation peuvent être dégagées qu'il faudra mettre au point.

<u>nomenclature (contenu)</u>: When two systems are exchanging data, following the cross correspondance of conceptual models, the content definitions have to be connected. Once more the problem can be solved by setting up correspondances between the definitions of objects attributes and links in

one system and a standardized coding system. The first draft of the CNIG catalogue is presently circulated among the participants of the commission in order to produce the first public draft by the end of summer 1990.

La définition d'une nomenclature d'objets géographiques susceptibles d'être échangés est une tâche importante. Différents organismes producteurs et/ou utilisateurs ont mis au point de tels catalogues pour leurs propres besoins et qui ont été rassemblés par le CNIG. Dans l'esprit de la séparation du particulier et du général, une nomenclature commune à tous les intervenants de la géographie doit être mise au point, ainsi que des nomenclatures propres à chaque domaine technique.

Il s'agit dans un premier temps de mettre au point une méthode de classement des objets des différents catalogues en fonction de leur importance sémantique relative et de leur emploi plus ou moins généralisé (structuration de connaissances).

Dans un second temps la méthode ainsi définie sera appliquée sur les catalogues recueillis par le CNIG, en séparant les objets en différentes catégories d'utilisation de façon à produire une nomenclature provisoire informatisée.

Proposer une méthode de comparaison des besoins en informations d'un organisme donné avec la nomenclature CNIG, de façon à affiner celle-ci en vue d'une mise au point définitive.

format d'échange: comparaison des propositions étrangères: Several countries have already produced standards or draft standards. It should have been inconsequent to ignore them. Among them NTF(UK), SDTS (US), Macdif (CDN), NESC (SA), and ATKIS (RFA) were taken into account. Within the road information world GDF is another standard to be considered. An international Working Group of nations belonging to NATO (DGIWG) is producing DIGEST as a digital geographic exchange standard.Considering the similarity of models which underlie the standards, one can define two families of standards. NTF and DIGEST have been chosen as the challengers for each family. The main conclusions of the experts group while comparing NTF and DIGEST with the aim of fulfilling the French requirements are: adoption of DIGEST as the provisional exchange standard for the French community.

Ainsi l'étude comparative de ces standards peut permettre d'appréhender les méthodes d'organisation, les modèles de données et les formats d'échange, et de déceler les pièges à éviter.

L'enjeu de la comparaison de NTF et de DIGEST, est d'établir, sur le plan institutionnel, théorique (modèle conceptuel de données), informatique (modèle de transmission), et organisationnel, laquelle des deux normes il serait préférable d'adopter pour remplir le cahier des charges précédemment fixé.

L'hypothèse du développement d'une norme française a été rapidement écarté eu égard aux aspects coût, délais et "européanisation" qu'une telle solution imposait.

La comparaison des aspects institutionnels de NTF et de DIGEST fait préférer la solution DIGEST en raison de sa dimension internationale, et particulièrement européenne, de la participation active de la France à son élaboration et à ses tests, et sa disponibilité imminente ne présente plus d obstacle.

L'aspect théorique de la comparaison porte sur l'adéquation du modèle conceptuel de données sous-jacent avec le modèle de données théorique "universel". Sur ce plan là, DIGEST s'est révélé plus performant d'autant plus que la documentation technique associée est apparue plus claire. Toutefois DIGEST ayant été créé pour des applications militaires à petite et moyenne échelle, il sera nécessaire d'enrichir sa nomenclature d'objets pour l'adapter à la diversité des applications civiles et de le tester sur des applications civiles à grande échelle.

L'intérêt d'avoir une norme reposant sur des standards existants a conduit ici encore à préférer DIGEST dans la mesure où le réemploi de la norme ISO8211 permet de laisser aux informaticiens les problèmes de transmission et de respecter une architecture en couches.

Le problème de l'échange de données géographiques se posant en France surtout en terme de transfert de fichier, l'adoption de ISO 8211 comme couche de transmission a été retenu. L'autre norme de transmission étudiée (ISO 9735-EDIFACT), plus adaptée à l'échange de message, a été écartée du fait de l'importance du travail à réaliser pour interfacer la couche géographique de DIGEST avec la norme 9735.

Le format DIGEST sous ISO8211 est le plus performant vis à vis des contraintes techniques et des obligations imposées par le cahier des charges. La commission propose donc de retenir ce format, d'enrichir sa nomenclature, de la tester, et d'attribuer à sa version civile le nom de **Echange de Données Informatisées GEOgraphiques (EDIGEO)**.

<u>format d échange: amorce de solution française</u>: The DIGEST choice for the French context implies a lot to do. This is a short term choice which implies to think to the future and the 1995 possible European standard. The aim is to prepare the use of DIGEST as the draft exchange standard for France. Turned to the name of EDIGEO, this standard supposes to be user-accessible. A testing program is in progress. Links have been made with other working groups dealing with geographic exchange (road navigation systems). The translation into French will be acheived during this summer.

La demande des utilisateurs d'informations géographiques numériques est relativement pressante et la préconisation par le CNIG de DIGEST doit pouvoir se traduire dans les fait dès 1991.

Ansi tester, répandre l'utilisation et préparer l'évolution de EDIGEO sont les tâches principales à accomplir.

Convaincre que EDIGEO convient, c'est réaliser des tests d'utilisation de cette norme montrant son adéquation aux usages projetés. Ils doivent impliquer les principaux intervenants. C'est aussi réaliser une francisation de la documentation. Basée sur un glossaire Franco-Anglais des termes et concepts tant théoriques que pratiques, sa traduction permettra d'être accessible aux utilisateurs futurs de la norme. C'est enfin de proposer d'ores et déjà des amendements à DIGEST pour l'améliorer.

L'accessibilité de la norme et son utilisation effective passe par la formation des utilisateurs, à différents niveaux de responsabilité.

C'est aussi accumuler et synthétiser une expérience nationale pouvant peser de tout son poids dans le cadre de travaux à l'échelle de l'Europe.

Convaincre que EDIGEO convient, c'est enfin montrer que, au delà des considérations purement théorique, son cadre permet d'intégrer des études antérieures et de fédérer les solutions déjà entrevues.

Le programme de test en cours de mise au point devraient être terminés avant la fin 1990, de façon à pouvoir proposer la version finale de la norme CNIG en Mars 1991.

Les contacts en cours avec les constructeurs automobiles (projet Prometheus et Drive) devraient permettre de faire une bonne synthèse entre les résultats de GDF et ceux de DIGEST, en jouant sur la complémentarité des études et en séparant nettement ce qui relève du domaine technique, de ce qui relève de la couche géographique et de ce qui relève de la couche de transmission.

<u>échange de messages</u>: As part of a study on Electronic Data Interchange (EDI) within a city, one can find the problem of exchanging spatially referrenced data. As the automatic exchange of messages between computers in a city are of interest for the future, adding capabilities for exchanging messages on geographic information is a real challenge. On experimental basis lying on real situations, the specific goal here is to create a set of messages which let it possible to exchange messages between GIS's.This new topic is part of the project known as EDI-Cité which is run under the auspices of EDI-France.

L'intérêt de l'échange de données et de la distribution interactive d'informations urbaines entre les collectivités territoriales et leurs partenaires publics et privés réside dans l'amélioration de la gestion en optimisant l'utilisation des informations urbaines. Il s'agit pour des situations d'échange prédéfinies (déclaration de travaux, permis de construire) d'automatiser les échanges de données informatiques.

La distribution interactive de données urbaines est permise par l'échange d'ordinateur à ordinateur, grâce à des moyens électroniques, de données structurées regroupées en messages normalisés. Reposant sur le modèle conceptuel de données "universel" implémenté dans EDIGEO, la normalisation des messages, et la gestion interactive de ces échanges, par la mise en réseau de SI urbains existants améliorera la diffusion d'informations géographiques à l'intérieur d'une ville.

CONCLUSION

As a conclusion the first table summerizes the starting point of French studies, i.e. which existing standard are kept, and the second one indicates the work to be done i.e. what are the main tascks to be undertaken to fulfill the requirements that were stated.

For each table the columns indicate the software layer. Transmission stands for the layer specialized in the physical transfer of data, geography stands for the layer where the organization of the data as geographic object is understandable for the computer and technical domain stands for the layer where the specificity of the domain let it possible to explain all of the meaning of every geographical information for the application (the consumption of the data for a specific purpose).

The lines are the four topics taking into account by the working group: Conceptual data model, content, file exchange, message exchange.

Table 1: What is to be re-used?

layer:	transmission	geography	technical domain
conceptual model		DIGEST	
content			(DEMETER)
exchange format	ISO8211	DIGEST	
message	ISO9735		

Table 2: What is to be done?

layer:	transmission	geography	technical domain
conceptual model		improve DIGEST Conceptual Data Model	specific rules for using the Conceptual Data Model
content		set up of the CNIG nomenclature	DEMETER nomencalture to be crossed with CNIG one
exchange format	get familiar with ISO8211	testing the use of DIGEST	upgrading of GDF to DIGES
message	get familiar with ISO9735	geographic message definition	use within EDI-Cité experime
	whom?	EDIGeo (EDI-France)	volunteer by domains

RAPPORTS DE LA COMMISSION

[Mars 90] Standard d'échange d'informations géographiques: rapport technique (114 pages) (french)

[Mars 90] Standard d'échange d'informations géographiques: rapport de synthèse de l'étude préalable (28 pages) (french and english)

[06-02-90]étude comparative des normes de transfert de données numériques ISO8211 et EDIFACT (32 pages) (french)

[Juin 90] rapport d'orientation (10 pages) (french)

Further details are available, please direct them to:

Monsieur Jean Denègre
CNIG
136 bis rue de Grenelle
75700 Paris
France

REFERENCES

American Congress on Surveying and Mapping (ACSM), 1988, "The proposed standard for digital cartographic data", The American Cartographer.

ALGRAIN, Ph, 1990, "Vers une normalisation des formats d'échange d'informations geographiques numériques", conférences MARI 90, Paris.

ATKIS, 1990, "ATKIS german specification", AdV.

BRÜGGEMANN, H, 1989, "Exchange Formats for Topographic-Cartographic Data-stage of discussion March 1989", surveying and Mapping Agency of Northrhine Westfalia.

CNIG, 1988a, "Mandat du groupe de travail normalisation des formats d'échange de données géographiques numériques, CNIG Paris.

CNIG, 1988b, "Journée Nationale de la Recherche Géographique", les actes de la journée du 13 Avril 1988.

CNIG, 1988c, la lettre du CNIG mai 1988/ N° 4.

CNIG, 1988d, la lettre du CNIG décembre 1988/ N° 5.

FICCDC, 1988, "a process for evaluating GIS", USGS open file report 88-105, Federal Interagency Coordinating Commitee on Digital Cartography-Technology Exchange Working Group.

FINGIS, 1989, "Cartographic Development and GIS activities in Finland", Finnish Proceeding of the 14th International Cartographic Conference ICA, Budapest, 12-24 August 1989.

JO, 1985, décret n°85-790 du 26 Juillet 1985 relatif au rôle et à la composition du Conseil National de l'information Géographique, journal officiel de la république Française.

NTF, 1988, National standard for transfer of Digital map data NTF, Ordnance Survey of Great Brittain .

SALGé, F, 1989a, le système d'information Géographiques National de l'IGN, VISUDA Paris 1989.

SALGé, F, 1989b, Geographic, cartographic, located data exchange formats: an overview of the existing solutions over the world, UDMS 31 Mai - 2 Juin 1989 Lisbonne.

SALGé, F, 1989c, "Conditions for dialogue between heterogeneous geographic information systems", OCDE Coordinated Information Systems for Urban Functioning and Management 11th-13th October 1989, Copenhagen (Denmark).

SALGé, F, 1990a, "l'échange et la diffusion des données géographiques numériques: synthèse et perspectives

Européennes", 3° journée de la recherche du CNIG Paris 22 Mai 1990.

SALGé, F, SCLAFER, M.N., 1990b, Interfacing GIS's, Euro Carto 8 Palma de majorque, avril 1990.

CLARKE, COOPER, LIEBENBERG, VAN ROOYEN "a National standard for the exchange of digital georeferenced information", National Research Institute for Mathematical Sciences of South Africa.

ACRONYMS USED IN THE TEXT

AFNOR	Association Française de NORmalisation french association for standardization
CAO/DAO	Conception/Dessin Assisté par Ordinateur computer aided design/manufacturing
CEN	Centre Européen de Normalisation european center for normalisation
CERCO	Club Européen des Responsables de la Cartographie Officielle European club of the official surveying agencies
CNIG	Conseil National de l Information Geographique national council for geographical information
DGI	information géographique numérique Digital Geographic Information
EDI	Echange de Données Informatisé Electronic Data Interchange
EEC	communauté économique européenne European Economic Community
ISO	organisation internationale de normalisation International Standard Organization
RNIS	Réseau Numérique á Intégration de Services integrated services digital network

GERMAN DEVELOPMENTS ON SPATIAL DATA EXCHANGE STANDARDS

Heinz Brüggemann
Surveying and Mapping Institute of Northrhine Westfalia
Muffendorfer Str. 19-21
D-5300 Bonn 2, FRG

ABSTRACT

The main current German standardization project in the field of digital cartography is the "Authoritative Topgographic Kartographic Information System (ATKIS)". The standard has been published in 1989. It consists of the three parts ATKIS Feature Class Catalog, ATKIS Symbol Catalog and ATKIS Data Model. The ATKIS developments are based on the German digital cadastre standards ALK for digital parcel registers and EDBS for spatial data exchange purposes. Further developments of European-wide spatial data exchange standards like ETF of CERCO are supported by German members in the special CERCO working groups.

INTRODUCTION

It is a legal task of the Surveying and Mapping Institutes of the German States to collect, document and provide basic land related data on the earth's surface. In addition, they have to produce, print, edit and distribute the official topographic maps and to look after the state's interests according to the use of maps and digital data by third parties. In the past this legal order lead to the result, that the user could rely on a supply of cartographic information covering the whole area of the state, precise and up-to-date according to well-accepted standards. The data have been provided in an analogue manner using maps. Maps and photo maps of different scales ensured, that for each application the needed data on the earth's surface were available.

The current realization of the German official basic information systems "Automatisierte Liegenschaftskarte (ALK)" and "Amtliches Topogra-phisch-Kartographisches Informations system (ATKIS)" i.e., Authoritative Topographic Kartographic Information System concerning information until now only available in topographic maps makes it necessary to find new ways of providing official surveying and mapping data. The requirements of the previous and future users of surveying and mapping data have to be observed. Especially, it has to be taken into account, that the users of these data prepare spatially related application oriented information systems, which have further to use the basic data of surveying and mapping, too. That means, that besides of feature classes, attributes, feature modelling rules and feature definitions particularly the needed link elements and link rules for the basic information system and the application oriented information systems have to be accomodated. Simultaneously common rules on the revision of the different information systems are to be fixed.

In the Federal Republic of Germany the "Arbeitsgemeinschaft der Vermessungsverwaltungen der Länder der Bundesrepublik Deutschland (AdV)" i.e., Working Committee of the Surveying and Mapping Institutes of the German States is responsible for the federal standardization activities. The development of exchange formats for topographic cartographic data started in the 70th. After the AdV had decided in 1972 on a standardized format for the exchange of cartographic data for special purposes of topographic mapping, in 1982 the "Einheitliche Datenbankschnittstelle (EDBS)" i.e., Uniform Database Interface has been published. This exchange format has been developed by the ALK project. It controls the data exchange with the ALK database. The most important current standardization project of the AdV is ATKIS. The ATKIS realization based on ALK components has almost been finished. At the present the EDBS is the ATKIS exchange format, too.

Recently the AdV installed a new experts group with the aim of revising the EDBS according to modern principles especially paying attention to European developments. AdV experts work intensively as members of working groups of the Comite Europeenne des Responsables de la Cartographie Officielle (CERCO) on the development of European standards, seeing that the growing international communication and the opening of the European market in 1993 make necessary international agreements on standards.

Compared with the activities of the AdV, other standardization efforts in the field of cartography are of minor importance.

SCOPE AND GENERAL GOALS

ATKIS Reference Model

When the AdV in 1985 asked an experts group to make a study

already existing ALK standard. But, soon it became clear, that a standard for large scaled cadastre information based on old hierarchical data model concepts could not be very well applied to cartographic medium and small scale applications. Whereas the cadastral map allows, because of it's scales between 1:500 and 1:2,000, to show the content of the cadastre on a map without problems, the presentation of the land features in scales 1:25,000 and smaller makes necessary partly extensive generalization efforts. But, generalization needs the complete knowledge of cartographic specialists, and it is not to be seen, that this working step can be replaced by a computer the next time. Therefore, it was obvious to describe the modelling of the landscape and the modelling of the map in different data models and to store them in different databases. So the Digital Land Models (DLM) contain the features on the earth's surface on their place even though simplified according to a defined accuracy. It is decided to build DLMs according to the map scales 1:25,000, 1:200,000 and 1:1,000,000 called DLM 25, DLM 200 and DLM 1,000. On the other hand, Digital Kartographic Models (DKM) derived from the DLMs contain only those features represented in the map to be realized. The limited place on the map make displacements, selections and simplifications necessary in parts with the result, that position and shape of the features represented by the DKM can be different from the real situation.

The basic separation of DLM and DKM leads to the ATKIS reference model shown in figure 1 containing the main data structures and processes as a rough scheme. Aspects of definition, of production and of communication are brought together.

The definition phase has been ended with the decision on the ATKIS documentation by the AdV in 1989. Now we have:

- o the ATKIS feature class catalogs concerning the DLM 25 and the DLM 200,
- o the specification of the DLM data model and the DKM data model,
- o the ATKIS symbol catalog concerning the DKM 25 resp. the map series 1:25,000.

The description of these documents is permanently being modified by AdV working groups because of the current data collection experiences. Feature Class Catalog, Data Model and Symbol Catalog are the basic components for the realization of ATKIS production systems by the different AdV governments. Based on suitable GISs like the ALK-GIAP of Northrhine Westfalia, the Siemens System SICAD or the Interraph System TIGRIS, the DLM and DKM data models are to be mapped on the system specific databases, and the data collecion procedures concerning the filling of the DLM; the generalization procedures to derive the DKM and presentation procedures to make analogue maps are to be implemented in a user friendly and efficient manner.

Map production cannot be the only goal of a topographic cartographic information system of the Surveying and Mapping Institutes, considering the upcoming use of modern digital techniques by the traditional users of our data, too. Already now, the demand for basic digital information on the earth's surface has become so big, that we have problems to satisfy all needs. With large efforts traditional users of our basic data, who until now used our maps, build their own application oriented information systems like road and traffic information systems, environmental protection information systems, planning information systems and soil information systems. They all rely on the basic data of surveying and mapping. However, they furthermore open the possibility of providing up-to-date application data to be integrated in ATKIS. During the last decade of this century we have to expect, that ATKIS will become part of a state-wide information network with extensive data exchange between different systems. So we came to the conclusion to take part in the important activities concerning common data exchange formats on both a national and an international level.

Vector and Raster Data
The ATKIS reference model (figure 1) shows, that we make differences between a vector based and a raster based DKM. The degree of abstraction of the DKM-V is very similar to that of the DLM; no symbolization has been done. However, the DKM-R contains all symbolization according to a special map design. The result is a pixel oriented data structure, which directly can be used to print the map from color separates. We think that it could be useful to provide both the DKM-V and the DKM-R to the user.

This assumption is supported by the fact, that a lot of users are asking for scanned map data to have background information for different applications. Therefore, it seems to be necessary to define not only standards for vector data exchange but also for raster data exchange.

National and International Standards
In view of the lot of already existing exchange format proposals dicussed on national and international levels, it makes no sense to find a very new standard. For the moment, the German EDBS is working and fulfills the main requirements. But, Germany is part of Europe and more and more applications don't stop at the borders of the European countries. Especially projects in the fields of environmental protection and of car navigation are acting European-wide and ask for commonly accepted data exchange formats.

The German Surveying and Mapping Institutes are confronted with three important standardization activities in Europe:

1) There is an old tradition concerning the cooperation of civil and militarian Surveying and Mapping Institutes in the most European countries. Therefore, the development of DIGEST by the NATO DGIWG group is an important basis for the discussion.

2) Since some years the car and electronic industries are preparing car navigation systems using digital maps. The Surveying and Mapping Institutes are interested in a close cooperation with these companies in providing the digital road data together with the road administration. The companies developed a specific standard called Geographic Data Files (GDF) based on the British NTF structure.

3) The most important activities are done by working groups of CERCO representing the interests of the surveying and mapping institutes of all countries belonging to the Council of Europe. The CERCO standard being developed is the European Transfer Format (ETF).

German representatives are involved in all of these developments. We especially support the CERCO activities and are willing to integrate the results of the CERCO discussions into the further development of the EDBS.

FROM ALK TO ATKIS

In 1970 the idea was borne to create a parcel register database in the responsibility of the cadastral administration on county level with direct link to the legal land registry in the responsibility of te court administration. The database should become the basis for a common land related information system existing of a lot of application oriented datasets and databases, in addition. In the following, the AdV decided to build first the "Automatisiertes Liegenschaftsbuch (ALB)" i.e., a database containing the describing part of the cadastre, and second the "Automatisierte Liegenschaftskarte (ALK)" i.e., a database containing the geometric/graphic part of the cadastre.

The ALB standardization activities started in 1970. The system development, based on available dataset standards, data access standards and programming languages standards, lead to an ALB pilot system in 1984 and to a production system in 1985. Now, nearly the whole country is covered by an ALB database.

In parallel, 1970 a second AdV working group was installed with the aim of developing the ALK part of the automated parcel cadastre. Because of missing conceptional and technical prerequisites, the activities were delayed until 1977. In this year a research and development project was set up funded by the German Federal Ministry of Research and Technology. The project started with a detailed conceptional phase, and the group came to the result, that it should be possible to develop a portable ALK system based on hierachical database interfacing standards and programming languages standards.

In the following, the main parts of the ALK, the ALK central database, the ALK-GIAP and the data exchange standard EDBS could be developed and later on introduced in 1986. It should be mentioned, that the ALK-GIAP was the first worth mentioning interactive graphic system using the new international standard GKS (Graphical Kernel System, ISO 7942) in the full functionality of level 2c.

The basic conceptual term of the ALK data model is the Object. An Object is the storage entity of a real world phenomenon described by it's geometric shape, by a classification number assigning it to the "Objektschlüsselkatalog (OSKA)" i.e., Feature Classification Catalog and partly by geometric and graphical elements needed in special cases for it's map symbolization. Because of the map scales 1:500 - 1:2,000 it is not necessary to make differences between the description of real world features and the features represented in a map. An important element of the data model is the external identification number, which allows the link to the ALB and to application oriented databases integrated in a common Land Information System.

Together with the legal land registry the German cadastre is the legal land property register and is based on the principle, that it has to be up-to-date at any time. Therefore, the updating procedure is based on legal rules, which lead to special database and interfacing techniques. The EDBS, which connects the ALK database part with the ALK data processing part is not only a passive data exchange standard, but also allows updating of existing databases on a command level integrated in the sequential EDBS structure.

ATKIS is based on the ALK developments. But, new information systems developments have been considered like object oriented data modelling approaches. The ATKIS standardization process started in 1985 and lead to the current ATKIS document accepted as a standard by the AdV in 1989. In parallel, the existing ALK software components, the ALK database system, the ALK-GIAP and the EDBS were developed considering the ATKIS requirements. Now in 1990 ATKIS implementations are available on the most graphic systems available in the field of cartography.

CURRENT EFFORTS

ATKIS Data Model

In spite of its federal structure the Federal Republic of Germany suceeded in designing and providing the official topographic map series by a common layout. Now we try to find a way of providing the digital spatial data to the different users in a clearly standardized manner. A main part of this standardization process is to come to a common understanding of the conceptual model to be used for the definition of the landscape features and the map featurers, for the storage of the DLM and the DKM data in an appropriate database system and for the derivation of exchange formats.

The main term of the data model is the Object (see figure 2: ATKIS DLM Structure). Features of the real world described and specified in the "Objektartenkatalog (ATKIS-OK)" i.e., Feature Class Catalog (see chapter 4.2) are stored in the DLM as DLM Objects. They consist of Object Parts of the types Point/Node, Line/Edge, Area/Ring or Raster and may be aggregated as Complex Objects. Attributes can be assigned to Complex Objects, Objects and Object Parts. The relation between an Object and it's Object Parts is 1:n, between Objects and Complex Objects are m:n relations allowed. Object Parts can be used to describe topological relations, too. They then become nodes, edges and rings of a graph. Overpassing relations are used to reference Object Parts above or below an Object Part.

The geometrical structure of an object is described by Geometrical Elements. m:n relationships between Object Parts and Geometrical Elements allow the description of line geometry without redundancy (planarity). Object Parts of Raster typ allow, together with their Geometrical Elements of Type Raster Matrix, the description of regularly ordered geometrical structures like point rasters of a digital terrain model.

The DKM data model contains in an analogue manner map features, which have been built from real world features by cartographic generalization. They are collected and classified in the "Signaturenkatalog (ATKIS-SK)" i.e., Symbol Catalog (see chapter 4.3). They don't carry any attributes.

The advantage of the ATKIS data model is, that it has been tailor-made for the structuring of land and map features. Therefore, it is possible to use a very similar terminology for digital data modelling and for application oriented definition and structuring of features in the ATKIS-OK and in the ATKIS-SK. During the ATKIS conception phase these circumstances simplified the discussion between data modellers and topographic and cartographic users.

A present problem of the ATKIS data model is, that its implementation using common available database management systems of hierarchical, network or relational type needs complicated tranformation steps. Better solutions based on object oriented database management systems are not available on the market.

A comparison of the object oriented structuring approach and the ATKIS data model shows, that

- o ATKIS Objects can be directly used as objects of the

object oriented data model (ODM),

- o ATKIS Feature Classes are identical with ODM object classes,

- o ODM type specialization is usable for type declaration of Object Parts as Point, Line, Area,

- o grouping can be used for Complex Objects,

- o local object classes of the ODM are needed to distinguish Feature Classes on Object level and on Complex Object level.

Special database developments like POSTGRES offer the possibility of describing the geometrical information in an adequate manner. Using such solutions, the ATKIS data model could be completely realized.

It would be very worthful to integrate some kind of behaviour of the features into the data model. The object oriented approach offers possibilities to describe e.g., constraints between attributes, objects and abject classes or rules for the DLM - DKM transfer process. Formal specification methods and languages are available and could be used to describe these aspects in a very clear manner as a basic for e.g., future object oriented data dictionaries. The result could be to get better standardization effects than now.

It should be mentioned, that such knowledge is already part of the ATKIS catalogs not yet described in a formal manner but by a lot of comments.

ATKIS Feature Class Catalog

The ATKIS Feature Class Catalog contains all topographic features characterizing the earth's surface. Feature Classes are grouped to Feature Groups, Feature Groups to Feature Categories. It is described, which attributes are to be assigned and which relations are to be set. Every feature class is in detail defined to declare exactly the modelling process for e.g., digitization processes. The rough structure of the catalog is shown in figure 3.

At the present the ATKIS-OK for the DLM 25 contains nearly 200 Feature Classes. The Feature Class definition has been done on the basis of the Feature and Attribute Coding Catalog (FACC) of the NATO. Other national and international proposals have been taken into account.

The use of the ATKIS-OK in practical work showed some problems, which we now try to eliminate. A problem is to specify the object modelling rules in such a way, that two different persons come to the same result, if they generate e.g., a piece of a road and assign the attributes on the right structuring level of this feature. We now add for every feature class a specific modeling description to guarantee the right data structuring.

ATKIS Symbol Catalog
Whereas the DLM shows the real world features completely with their original position, the DKM contains only those features, which find a place on the designated map. In addition it can happen, that features have to be more simplified and to be displaced. Therefore, it is not convenient to use the DKM data for quantitative analyses. It is the main result of cartographic generalization work and can be used to produce automatically different kinds of maps.

The ATKIS-SK (figure 4) has to fulfill two main tasks:

1) It has to describe all Feature Classes, wich should be shown on the designated map series. In addition it has to fix all rules needed to derive DKM features from DLM features. DKM features do not have attributes. So the DLM Feature Classes had to be splitted into a lot of more DKM Feature Classes. The reason is why we don't have so many selection requirements on DKM Objects as on DLM Objects.

2) As its second task the ATKIS Symbol Catalog has to describe exactly the graphical symbolization of the map. In addition there are precise rules for the automatic symbolization procedure like priorities of map symbols and parts of them in case of overlapping, shielding rules and rules for repeating graphical structures. Important is the fact, that the space needed by a symbol is the main constraint for the generalization process. Therefore, the ATKIS-SK controls this process, too.

Symbol catalogs are digital symbolization standards. In an ideal manner they could contain both the cartographic knowledge on the generalization process and the cartographic knowledge on the map symbolization process. In similar manner as the ATKIS-OK the ATKIS-SK could be specified by object oriented methods. Doing so there could be goud conditions concerning the transformation of the DLM into the DKM and concerning the standardization process of map symbolization and map production. Exchange formats should offer possibilities of transfering both the Feature Class Catalog and the Symbol Catalog for application purposes by the user.

Already now it has become apparent, that deriving automatically a DKM from a DLM makes necessary to add further elements to the existing standards ATKIS-OK, ATKIS-SK and ATKIS Data Model. Especially the cartographic knowledge on the processes needed for the realization of the whole ATKIS production system (high sophisticated image analysis for the DLM revision, cartographic generalization, deriving map products in an automatic way) should be described by formal specification methods and integrated so far as possible into the ATKIS standard.

The EDBS
The EDBS is the current AdV standard in the field of spatial data exchange. It has been developed as part of the ALK standard. It is not only a passive exchange structure but contains also commands for changing existing databases because of the following requirements:

- o Taking over digital data into the databases
- o Communication between primary database and secundary database,
- o Providing data to the users.

To fulfill these requirements the EDBS has to support the following functions:

- o First filling of the database,
- o Revision of the existing database,
- o Selection of data out of the database,
- o Providing data to the users,
- o Updating secondary databases,
- o Creation of a process protocol,
- o Functions of order organisation.

Based on the ALK EDBS we have developed an ATKIS EDBS supporting the special ATKIS data struture. Both EDBS solutions are very close to the ALK/ATKIS logical data structure used in a specific hierarchically structured database solution. It is not suitable for an international discussion. But, the dynamic approach of this standard concerning the integration of commands changing existing databases is worth to be discussed on an international level, too. From our point of view, the communication between basic spatial information systems and application oriented spatial information systems will become an important job for the last decade of our century.

Future Work
The first we have to do is to control the practicability of the developed ATKIS standard based on the first experiences we have. The AdV has just installed new working groups to complete and correct the three parts of the standard, the ATKIS-OK, the ATKIS-SK and the ATKIS Data Model. The special working group concerned with the data model has to find out the best way of defining a new EDBS version based on modern principles. It is not our intention to develop a new standard; the present idea is to find a common understanding on a European level based on the cooperation within the CERCO

working groups.

A second main emphasis will be a modelling standard for real features stored in the DLM and in the DKM. This activity has to ensure, that the features stored in different databases of different States in Germany are built on the same principles. Otherwise communication could become very complicated on a semantic level.

A third activity has been started to find out the future map symbolization of our official topographic maps. This work will lead to new symbol catalogs based on new knowledge of graphic information theory and new map production possibilities.

REFERENCES

Arbeitsgemeinschaft der Vermessungsvserwaltungen der Länder der Bun-desrepublik Deutschland, 1975, 1984, "Sollkonzept Automatisiertes Liegenschaftska-taster als Basis der Grundstücksdatenbank" Band 2: Automatisierte Liegenschaftskarte, Hannover 1975, Logische Datenstruktur Grundrißdatei, Hannover 1984.

Arbeitsgemeinschaft der Vermessungsvserwaltungen der Länder der Bun-desrepublik Deutschland, 1982, "Einheitliche Datenbankschnittstelle (EDBS)", Hannover.

Arbeitsgemeinschaft der Vermessungsvserwaltungen der Länder der Bun-desrepublik Deutschland, 1989, "Amtliches Topographisch-Kartographisches Informationssystem (ATKIS)", Bonn.

Brüggemann, Heinz, 1987, "Standardisierungsbemühungen im Bereich raum-bezogener Anwendungen", GKS-Tage, Essen (Tagungsband).

Brüggemann, Heinz, 1988, "Brauchen wir Standards für AM/FM-Systeme?", AM/FM-Regionalkonferenz Siegen, (Tagungsband).

Digital Cartographic Data Standards Task Force, 1988, "The Proposed Standard for Digital Cartographic Data", dedicated issue of the American Cartographer, 15(1).

Digital Geographic Information Working Group, 1989, "Digital Geographic Information Exchange Standards (DIGEST)". "Geographic Data Files Standard (GDF)", 1988, proposed by Philips International BV and Robert Bosch GmbH, Eindhoven.

ISO 8211, 1986, "Specification for a Data Descriptive File for Information Interchange".

Stonebraker, M. and L. Rowe, 1986, "The Design of POSTGRES", in Proc. of ACM SIGMOD Conference on Management of Data, Washington D.C..

Working Party to Produce National Standards for the Transfer of Digital Map Data, 1986, "National Transfer Format (NTF)", Southampton.

Official reports on ATKIS are available at:

Landesvermessungsamt Nordrhein-Westfalen
Muffendorfer Str. 19-21
D-5300 Bonn 2

Official reports on ALK and EDBS are available at:

Niedersächsisches Landesverwaltungsamt
- Landesvermessung -
Warmbüchenkamp 2
D-3000 Hannover 1

STANDARDIZATION EFFORTS IN HUNGARY

Pál Divényi
Institute of Geodesy, Cartography and Remote Sensing
Guszev u. 19., 1051 Budapest, Hungary

INTRODUCTION

Since the introduction of computer sciences in cartography, the digital storing and application of mapping data have been showing more and more complicated structures and gaining new users. The software, oriented to handle map-based information, however, requires various base-maps (or background maps) of different scales, information and accuracy. The object of map use determines the structure of data storage, the portability of programs and data, and the accuracy of data. The method of representation depends on the user environment, the need for accuracy of representation and information, the financial resources of customers, the technological background and the man-machine relationship.

Naturally, the wide range of digital information supplied by several sources can be most effectively and most widely used, if we have an established, standardized way of data communication, which is able to link various computer systems and to preserve the original information as well.

BACKGROUND

The need for a standardized data transfer was already born towards the end of the 1970's in countries that made great achievements in informatics. In the United States, the American Congress on Surveying and Mapping (ACSM) established the National Committee for Digital Cartographic Data Standards, which published its proposed standards in the January 1988 issue of The American Cartographer. The National Transfer Format (NTF), a proposal for Standards, appeared in January 1987 in Great Britain. In West Germany, it was decided in 1986 to introduce Amtliche Topographisch-Kartographische Informationssystem (ATKIS) as a national standard. The standard format was prepared by Arbeitsgemeinschaft der Vermessungsverwaltungen der Länder der BRD (AdV) in 1988.

In Hungary, the increasing production of digital cartographic data has been observable since the second half of the 1980's. The need for a data exchange standard in Hungary was raised by the parallel work in several institutions. The Cartographic Department of Eötvös Loránd University, Budapest, organized a working group with the cooperation of experts from several agencies in order to prepare a proposal for standards. First, the Canadian, American, British and German issues were studied together with a lot of publications in the topic. The working group made it clear on the basis of their and foreign experience that several major aspects must be considered in the work.

REQUIREMENTS

- The data transfer format should be as simple as possible.
- The size of transferred data in the format should be as small as possible.
- It is advisable that the receiver should be able to define the logical and physical structure of data without using other documentation.
- The data transfer format must be usable for vector data and raster data exchange as well.
- The standard must contain the definitions of terms.
- The professional standards must be used in the management of different data types.
- The data transfer format must be independent of the data carrier.
- The data transfer format must not depend on the type of computer.
- The format must be ready to input new objects and terms.
- The opportunity of transferring information on changes needed by updating the database must be facilitated.
- The standards should contain methods for error spotting and data control.
- The transfer of data of non-contoured maps must also be available.
- The problems associated with dynamic maps must be solved.

It became obvious that none of the studied issues were directly usable in Hungary; they were all prepared in countries that already had a kind of cartographic databank, and the standard reflected its format. It would have been unwise to accept any of these issues, because they depend on hardware, and they record the state of development at the time of their establishment.

SCOPE AND GENERAL GOALS

According to the above considerations, the working Group has proposed the following as basic guidelines:

- o The working group accepts the sheets of the Uniform National Map System as a base and accepts all its specifications that have already been standardized (e.g. legend).
- o Proposals for the standard will be elaborated for three various scales: 1:10,000, 1:100,000, and 1:200,000 or 1: 300,000.
- o The new data transfer standard will be thoroughly used in the profession.

The execution of the above issues means the following tasks:

- o list the terms and compile a dictionary that explains the major expressions,
- o discuss the problem of data quality,
- o describe the classification of map information, and
- o review the data transfer structure and record description of the British standard (NTF) in an annex, which with certain modifications might serve as a basis for the Hungarian system.

CONCEPTUAL PLATFORM

First, the basic terms should be defined.

- o /map/ ELEMENT: the object referring to identical types of features of the real world that can not be divided further in the given map representation.
- o /digital map/ OBJECT: the digital representation of a map element.

In order to facilitate the digital representation of map elements or the exact coding of objects and their efficient operations, the objects must be defined in details and in general as well. The permitted 0-, 1- and 2-dimensional map objects are divided into three groups.

- o Pure geometric objects /G/, which only help drawing maps.
- o Geometric and topological objects /GT/, which serve as a basis for modern mapping operations.
- o Pure topological objects /T/, which are the basis for certain analyses.

There are special objects /X/ as well, which are built on points and chains.

Further on, the definitions of simple map objects follow. A simple map object can not be divided further. The definitions are applicable for plain (e.g. Euclidean) and spherical or ellipsoidal surfaces. Complex objects are built from simple objects belonging to the same type of geometry /G/ or topology /T/. In accordance with this standard, every complex map object (e.g. defined by the user) must be built up from the above simple objects.

DEFINITIONS

To summarize the terminology of the standard is a major task. This part of the paper only lists the Hungarian equivalents of English terms in the dictionary from A to D.

ABSOLUTE CO-ORDINATES	Abszolút Koordináták
ACCURACY	Pontosság
ADDRESSABLE SPACE	Címezehtó Tér
ANALOG	Analóg
ARC	ÍV
AREA	Terúlet
AREA SEED	Terúleti Jellemzó
ATTRIBUTE	Jellemzó
ATTRIBUTE CODE	Jellemzó Kód
AUTOMATIC DIGITISING	Autommatkus Digitalizálás
BINARY	Bináris
BINARY DIGIT	Bináris Számkegy
BIT	Bit

BLIND DIGITISING	Vak Digitalizálás
BLOCK	Blokk
BYTE	Bájt
CARTOGRAPHIC DATABANK	Térképészeti Adatbank
CHAIN	Lánc
CHAIN ENCODING	Láncolás
CHARACTER CODE	Karakterkód
CHARACTER SET	Karakterkészlet
CHARACTER STRING	Karakterlánc
COMPACTION	Tömörítés
COMPUTER ASSISTED CARTOGRAPHY	Számítógépes Térképészet
CURSOR	Kurzor
CURVE FITTING	Görbe Illesztés
DATABANK	Adatbank
DATABASE	Adatbázis
DATABASE MANAGEMENT SYSTEM (DBMS)	Adatbáziskezeló Rendszer
DATA CAPTURE	Közvetlen Adatnyerés
DATA ELEMENT	Elemi Adat
DATA MANAGEMENT	Adatkezelés
DATA MODEL	Adatmodell
DATA SET	Adathalmaz
DATA STRUCTURE	Adatstruktúra
DATA TYPE	Adattípus
DENSITY	Súrúség
DIGITAL ELEVATION MODEL	Digitális Terepmodell
DIGITAL GROUND MODEL	Digitális Domborzatmodell
DIGITAL MAP DATA	Digitális Térképadatok

DIGITISING	Digitalizálás
DIGITISING TABLE	Digitalizáló Tábla

ENCODING OF TOPOGRAPHIC OBJECTS

This chapter describes the classification of topographic objects of the Uniform National Map System for topographic maps at 1: 10,000, 1:25,000 and 1:100,000. This classification is intended to help the development the standard for computer processing of the data of objects. The work has been basically carried out according to the regulation, i.e. the legend specifications for the topographic map series of the above scales by the Office of Lands and Mapping, Ministry of Agriculture and Food. In the classification of topographic objects the working group has considered various foreign digital standards: the Proposed Standard for Digital Cartographic Data - USA, National Transfer Format - Great Britain, Classification Code of Topographic Information by the Geodetic Service of the USSR and coordinated with the geodetic services of East European countries and also considered digital information systems Amtliche Topographisch-Kartographische Informations system - (ATKIS) FRG/.

The classification covers the objects, the attributes of objects and their codes. The objects are classified in a hierarchic order. On the first level, the objects are divided into 9 classes, which - with some purposeful modifications - follow the categories of the legend of the Uniform National Map System. On the second level, the objects are thematically grouped in order to ease orientation. The objects can be found on the third level. The number of elements is never more than 10 in a group. The method that those objects are marked with an asterisk (*) that are only represented on 1:10,000 topographic maps already shows the tendency of automation in generalization. The number of objects is 207, 24 of which are marked with an asterisk. In the table, the name of the object is followed with its identification number in the legend.

The classes are referred to by mnemonic letter codes; the second and third levels of classification are encoded with one digit respectively.

Every object has its specific attributes. The attributes are listed in an alphabetical order and they number 75. An example is presented in Figure 1.

Class	Objects	Attributes
1.1	Control points /A/	
01	Astronomical control point /1/	16, 36, 45, 59
02	Triangulation point /2/	01, 07, 24, 29, 32, 36, 45, 47, 49, 59
03*	Orientation point /azimuth/ or reference point, boundary mark /6/	24, 49
04	Traverse point /7/	01, 07, 24, 29, 32, 47, 49, 59
05	Bench mark /8, 9/	01, 07, 24, 29, 36, 45, 59, 72
06	Control points of vertical crustal movements /10/	01, 49
07	Spot height /272, 273/	01, 32, 72

Figure 1. The classification of topographic objects.

DATA QUALITY

When dealing with the standardization of data quality, two aspects must be considered: the evaluation of data quality and the transfer of qualitative information.

The producers of data must have certain prescriptions for their method, or otherwise it is impossible to define the terminology that describes quality. /A statement that quality codes are "99% correct" raises the question whether a file which is "99% correct" can be directly compared to an other "99% correct" file./

The evaluation of quality can only be circumstantially expressed with numbers; most of the data producers are unable to supply decision-makers with the necessary amount of samples. Therefore, some suggestions on the evaluation of quality are only reviewed here.

Concerning the transfer of qualitative information, the following should be emphasized.

- o Some of the information can be expressed by numbers, while others need a description by words.

- o Some of the information are conveyed into the database as data, while others are normally referred in data transfer only.

These differences are usually not clear and they may change in different& data transfer processes. The future standard must definitely clarify this question as well. The following aspects are proposed to be considered in the evaluation of data quality and data transfer.

- o Information must be available on the accuracy of positioning, accuracy depending on attributes, state of maintenance, course of processing' completeness and method of data acquisition.

- o Information must be easily accessible, particularly the "physical" transfer information.

- o The representation method of qualitative data must be flexible, clear and understandable. The information has to be displayed at various levels, such as the database level, partial (map) level and attributes level.

- o Where possible, the information must be specified by numbers in order to be able to compare them to other files.

- o The information must be independent of scale, that is the positional accuracy must be expressed in terms of distance on the reference surface.

The general approach to the problem of data quality is, first of all, conditioned by an agreement between the user and supplier, this responsibility can not be shifted upon the user. If it is suggested that the problem of data quality is an important question and that the transfer of qualitative data is insured, the data suppliers may be encouraged to use the suggested opportunities. It is important that the receivers of the data easily recognize any error in the quality of data.

Problems associated with accuracy, generalization and representation of map data arise in every both positional and contextual aspect of data collecting.

The data are carried through various phases of data collection processes (e.g. digitizing) before they reach the user. These phases often weaken their quality, although there are errors that can be detected and corrected.

There are two ways of giving the final evaluation on the quality of data sets.

- Checking samples are taken from the original and the final product in order to evaluate accuracy. In the case of topographic mapping, this process means field-checking and additional observations. This is the best and most reliable method.

- The effect of devices and procedures on accuracy is calculated for the different phases of processing, and these estimates are used to estimate the accuracy of the final product. This method is less reliable, but it is the only possible way very often.

The numerical expression of data quality is without any meaning, if the way the information was acquired is unknown. This means that the compact way of transferring information on data quality is not only complicated, but also dangerous.

In accordance with the above considerations, it was decided to produce a short historical documentation or description on every file, to explain the acquisition of results, and to describe the controls made over data, the method or both. The proposed content of a documentation like this will be described later.

Particular attention should be paid to positional accuracy, attributes accuracy and to problems associated with time. Some indirect (but less reliable) comparisons can be made between data from different origin, but it is much more important to evaluate data of identical origin.

The proposed standard discusses the other factors of data quality on several pages.

STANDARDIZATION OF DIGITAL CADASTRAL MAPPING

The digital cadastral base-map is the digital form of the cadastral base-map of the Uniform National Map System designed for computer processing and use; it is a communicational data system; its data content and the reliability of data satisfy the demands prescribed in the regulations for cadastral base-maps, while its structure allows the interchangeable use of cadastral basic data in digital form. The digital cadastral map (DCM) is represented on data carriers in the form of physical files.

Organizations (firms, undertakings) producing cadastral and mapping data and government agencies distributing data (Institute of Geodesy, Cartography and Remote Sensing, land offices) are obliged to produce and distribute the DCM according to the data system described in the present regulation, if they have the necessary computer facilities.

The DCM also describes the logical relationships between the geometric elements of the cadastral base-map, namely, it serves the topological data model of the base-map for further processing. The data system is independent of the map sheet system.

The content of the DCM is based on the national guidelines for cadastral base mapping. In addition, the DCM also includes geocodes according to a decree of the Ministry of Agriculture and Food.

The data in the DCM contain information for complete parcels only. The DCM has no data referring to the sheet system. The information corresponding to texts and figures outside the map frame must be partially or completely supplied from the DCM. The structure of data storage allows the selective handling of types of objects, single objects or objects according to their characteristic data.

The Office of Lands and Mapping of the Ministry of Agriculture and Food may permit the construction of the DCM not in harmony with the content of the national guidelines (e.g. without parcel content, public place or relief etc.).

The geometric elements of the map content are included in the DCM as spatial objects.

The objects of the DCM are to be identified by the computer in two ways:

- o geometric identification with the use of geocodes,
- o alphanumerical identification with the use of numbers and/or codes according to the present or referred guidelines.

The data in the DCM are divided into two groups:

- o planimetric information,
- o relief information.

The data of objects must be compiled from their planimetric data (geometric description) and from their individual geometric and alphanumerical identifiers.

The inscriptions within the sheet frame are linked to the planimetric features or relief objects as their textual information. The types of objects and their names in the DCM are presented in Figure 2.

Object	Letter code	Feature code
Administrative unit	K	2E
District	B	2C
Category	O	2O
Block	T	2B
Parcel	F	2A
Sub-parcel	A	2a
Building	E	2e
Line/point	V	1v
Cadastral control point	P	Op

Figure 2. Types of objects and representative codes.

The identifiers and names of objects, their linking and other data must be built as physical files in the DCM. The objects in the DCM are enframed and/or assembled objects. The basic object is a parcel. Enframed objects generally have area, they are of the same type, and they are situated as mosaics on the larger area of an other object. They can also be objects of the same type without size (points) and placed on the surface of an object or objects having area.

Assembled objects are defined as objects that have an enframed object or objects. Since an enframed object can be an assembled object too, there is a hierarchic relationship between the objects which is expressed by geocodes in the DCM. Every object (or data linked to objects) may be either provisional or final. Provisional data must also be organized into a communicational data system on the basis of the present guidelines of the DCM. The definitions and explanatory texts of objects to be stored in the DCM are included in the Annex.

TECHNICAL REPORTS ON DATA STANDARDIZATION

Digitális Térképi adatok átviteli szabványa, 1989, Tervezet. (Interchange standard for digital maps. Proposal.) Szerk.: ad hoc munkabizottság (ELTE Térképtud. Tanszék es ászsz). Ed.: working group. Budapest, Studia Cartologica, Vol. 11, p. 50.

Szabályzat a digitális földmérési alaptérkép szabvávyosításárá, 1989, Tervezet. (Data interchange standard for cadastral mapping. Proposal.) Ed.: Institute of Geodesy, Cartography and Remote Sensing. Budapest, p. 53.

REFERENCES

Amtliche Topographisch-Kartographische Informationssystem, ATKIS, 1988, Vertrieb Landvermessungsamt NRW.

Digital Cartographic Data Standards Task Force, 1988, "The Proposed Standard for Digital Cartographic Data", The American Cartographer, 15 (1), p. 140.

International Standards Organization, 1985, Information Processing: Specification for a Data Descriptive File for Information Interchange, ISO 8211.

Moellering, H., 1988, "Fundamental Concepts that Form the Basis of the Proposed American cartographic Data Exchange Standard, Cartography, 17 (2), p. 9-14.

Morrison, J.L., 1988, "Digital Cartographic Data Standards: The U.S. Experience", Cartography, 17 (2), p. 1-7.

Objektartenkatalog, ATKIS-OK, 1988, Vertrieb Landesvermessungsamt NRW.

Térbeli információrendszerek kialakítása, 1980, Tanulmány. Study on the development of GIS/BME Geod. Int., Budapest, p. 130.

US Geological Survey, Digital Cartographic Data Standards:
Digital Line Graphs from 1:24,000-scale Maps Digital Line Graphs from 1:2,000,000-scale Maps
Geographic Name Information System
Digital Elevation Models

Working Party to produce National Standards for the Transfer of Digital Map Data, 1986, "Final Draft Papers", Great Britain, p. 85.

THE STANDARDIZED EFFORTS FOR DIGITAL CARTOGRAPH DATA IN JAPAN

Minoru Akuyama
Map Information Manager

Akira Yauchi
Head of Survey Guidance Division

Geographical Survey Institute
Kitasato-1
Tsukuba-shi
Ibaraki-ken

INTRODUCTION

A set of digital cartographic data covering all over Japan, which named Digital National Land Information, was first created in 1974. Since then, several sets of digital cartographic data have been created independently for different purposes.

Currently standardization efforts have been made in each application field, and no single general purpose standards have yet been proposed. There are four major application fields of digital cartographic data in Japan. They are utility facility management corresponding to 1:2,500 scale maps, vehicle navigation around 1:25,000, and national or regional planning and analysis with 1:200,000 and smaller scales.

System users of those cartographic databases do not seemed to expect the collaborative use of different scale cartographic data into their own application. They are rather likely to insist on their own data structure, data items, and data format for the optimal use of their own system. However, standardization within each of the above mentioned four fields has been practically accomplished to some extent.

As for utility facility management, the **Road Aministration Information Center** is the leading organization to create the digital cartographic data of major cities in Japan. Therefore, its specification practically acts as the standards of this application field.

For urban planning, municipalities are the main bodies to prepare digital cartographic data. In order to avoid that wide variety of data specifications are getting in use in this field, the **Geographical Survey Institute** established the **Standard Procedure and Data Format for Digital Mapping** for the purpose of advisory and guiding administration to local governments.

Vehicle navigation system has not yet been so popular in Japan, however, Toyota, Nissan (Datson), and Mazda already sell cars with navigation systems and other automobile companies and car equipment manfacturers are preparing to supply similar systems. Currently the **Japan Digital Road Map Association** is the only organization to create digital road map data for car navigation systems. And all the aforementioned three automobile companies have been supplied with digital road map data from the Japan Digital Road Map Association.

For small scale cartographic data, the **Geographical Survey Institute** has prepared at the very early stage and those data have been used widely. Therefore, the specification of those data is recognized to be the standard in this field.

Besides those widely used digital cartographic data, there exist some other sets of data used in limied application fields. Some of them are the **Digital Cadastral Map Information** of the **National Land Agency**, the **House Directory Map Information** of the **Zen-rin** corporation, and the **Detailed Digital Land Use Data** of the **Geographical Survey Institute.**

SCOPE AND GENERAL GOALS OF THE STANDARDS EFFORT

There exist no single general purpose standards for digital cartographic data exchange and effort for such standards has not yet been active in Japan so far. This seems to mean that we are in chaos and there is no hope to change the situation in the near future. However, we are quite optimistic to this situation and for the future.

Data files can be classified into three types. They are internal format of data creation systems, data exchange format, and internal format of a user's system. If no data transfer to other systems is expected, it is the most effective and natural to create data in user system format.

In this regard, digital cartographic data within the limited application fields does not need to be standardized. Strict standardization may suppress probable progress and improvement of technology. Therefore, it is considered that standardization effort should be limited to those data which are expected to be used in various purposes and transferred to various users. Moreover, expectations to standards vary with major application fields and users, and thus associated map scales.

The optimal standards for general purpose digital cartographic data throughout scales might be inconvenient and redundant to each specific fields. Therefore, at least three different standards corresponding to large, medium, and small scale digital cartographic data are supposed to be necessary as the general goals.

Currently medium and small scale cartographic data creation and utilization are rather limited and the desire for such standards is not so urgent. Confusion would reside in the urban planning application field. For this purpose, digital cartographic data of 1:2,500 or larger scale topographic maps are to be created by huge numbers of municipalities as 3,300 and utilized in various subsystems of municipal administration. Furthermore, those data might be supplied to the public as traditional graphic maps are.

Under this situation, standardization for this type of data was recognized to be essential. Then the **Geographical Survey Institute** formed the **Committee for Digital Mapping Standardization** in 1985, and after a three-year study, the **Standard Procedure and Data Format for Digital Mapping** was established in 1988. This standard include standard procedure and accuracy limits to assure standard data quality, standard code system to cover general large scale cartographic entities and data format to assure smooth data transfer between different systems.

This standard is expected to promote an adoption of urban information management techniques, development of high quality urban information system softwares, as well as digital data preparation and maintenance.

BRIEF HISTORY AND BACKGROUND

Digital National Land Information (Miyazaki and Tsukahara, 1987)
In 1974, the **Geographical Survey Institute** started the project of preparing the **Digital National Land Information** and have been collecting various kinds of nationwide digital information concerning the national land and human activities in cooperation with the **National Land Agency.** This is the first digital cartographic data covering all over Japan.

The coordinate system of the data is the standard regional grid system which is based on longitude and latitude. The whole land of Japan is divided into more than 100 primary grids of one degree longitude by forty minute latitude. The secondary grid is defined as dividing the primary grid into 8x8 portions. And the tertiary grid is 1/10 section of the secondary grid along lines of longitude and latitude. The area of the tertiary grid is approximately one square kilometer in the central Japan, while it various in accordance with latitude. Point coordinate is described by the normalized coordinates within the corresponding secondary grid.

The contents of the Digital National Land Information are ranging from topographic features to social economic statistics. Each item data was created independently in different years without any intention of standardization. Currently those sets of data have been revised repeatedly.

Through the revision work, data formats have been reconsidered to be unified to some extent.

Detailed Digital Land Use Data (Otake, et. al., 1987)
In order to grasp trend of housing land use in three megolopolitan areas of Tokyo, Osaka, and Nagoya, the **Geographic Survey Institute** started the project to prepare digital information on housing land use situation of every five years in 1981. The sets of data first created are as of 1974 and 1979. Then data sets of 1984 and 1989 have been prepared afterwards.

The structure of the set of data is raster form of 10m square and 100m square pixels. The number of land use categories is fifteen. And image-like data set of 3km by 4km are is stored as a unit in the data file.

Digital Cartographic Data from 1:25,000 Scale Maps (G.S.I., 1987)
In 1984 the **Geographical Survey Institute** started another digital map information collection project and has been collecting the data for fundamental categories of 1:25,000 scale topographic maps, namely contour lines, political boundaries, roads, railways, and water edge lines.

The objective of the project is to prepare a ditigal cartographic database from 1:25,000 scale maps for the purpose of automated map compilation and production as well as satisfying public demand.

In spite of the initial plan of the project, the pace of the project execution is very slow. Currently only political boundary data and a part of contour line data have been digitized. Because of the incompletion of data preparation, the collected data have not yet been open to public and only used inhouse. Therefore, standardization as not yet been considered to this set of data.

Road Administration Information System (Tada, et. al., 1990)
The **Road Administration Information System (ROADIS)** was designed to provide information on roads and utilities to road administrators as well as public utility organizations. In order to execute the said project, the **Road Administration Information Center (ROADIC)** a foundation organized under the sponsorship of the combined organizations of the subject municipalities, public utility organizations, and the Ministry of Construction was established in 1985.

ROADIC began the developmment of the system in 1985 under the supervision of the **ROADIS Development Committee** which was chaired by Professor Masao Iri of the University of Tokyo and was composed of Metropolitan Government and the Ministry of Construction, representatives from the Police Agency, and various public utility organizations, i.e., telecommunication, electric power, gas, water supply, sewerage, subway, etc., and ROADIC.

Under the ROADIS Development Committee, a task force was established for the development of ROADIS. The members of the task force were composed of specialists in road administration, traffic control, public utilities, computer science, cartography, and surveying. The database have been built by ROADIC and Database Building Joint Ventures, as well as by utility companies.

Pilot Projects were started in 1987 in Yokohama and Kawasaki, both located adjacent to Tokyo. Currently, preparation of system building was started in other cities. The experimental results from the pilot project were reflected through revisions to system developmment during the implementation stages. Trial operational usage of the system is expected to start in 1990 in the pilot project areas.

After examination and evaluation of existing Road Register Maps and Utility Maps, the **Standard Specification for Building ROADIS Databases** was established. Through the practical use of the database the standard has been revised continuously.

1:2,500 General Purpose Digital Map Data Base Standard
In accordance with the City Planning Law, each municipalities should make and revise landuse plan for urban planning district on 1:2,500 topographic maps every five years. Currently urban information management techniques have been introduced to this field, and, therefore, digital cartographic data corresponding to 1:2,500 topographic maps have been created in several municipalities.

Different from other application fields of digital cartographic data, large numbers of organizations are involved in data creation and utilization. Moreover, since those topographic maps have been sold, in most cases, as large scale base maps, digital cartographic data of those maps are also expected to be supplied to other organizations. Therefore, standardization to assure exchangeability is considered to be necessary in this field.

In this regard, **1:2,500 General Purpose Digital Map Data Base Standard** was established in 1986 by the **Ministry of Construction.** In 1985, the **Promoting Headquarters of 1:2,500 General Purpose Digital Map Data Base Preparation** was established in the **Geograpahical Survey Institute.** The **Task Force for Standards** of the group wrote a draft standard which was examined and authorized by the **Committee for**

Standardization chaired by Professor Masao Iri of the University of Tokyo and composed of scholars, officials of the Ministry of Constructaion, and others.

In spite of dense discussion and experiments carried out before determining the final standards, it has not widely been adopted. The reason may reside in balance of cost and expections. In the standard, more attention was paid on complete exchangeability than flexibility to meet user requirements. Afterwards, this standard was found to be unable to act as standard and replaced by another standard.

Standard Procedure and Data Format for Digital Mapping (Hishino and DInaba, 1987), (Akiyama. er. al., 1988)

In order to take full advantage of this technology and to bring it into wide use it is essential to standardize the procedure and data format. In this regard, the **Geographical Survey Institute** gathered the most of all Japanese aero-survey companies involved/interested in research and business of digital mapping technology to establish Japanese standard in 1984. In the meantime, the Geographical Survey Institute set up by the **Committee for Digital Mapping Standardization** chaired by Mr. Eiji Inoue, vice-president of the Japanese Association of Surveyors, and composed of scholars and experts in the field of photogrammetry, computer mapping, and municipal administration to discuss and investigate the standard.

Through a one-year preparatory study and a three-year investigation, the Geographical Survey Institute and the committee established the **Standard Procedure and Data Format for Digital Mapping** in March 1988. The purpose of the standard is to guarantee data accuracy and compatibility so that the data can be utilized by other parties as well.

This standard was expected to be adopted only for "Digital Mapping," however, the exchange format determined in this standard is now used for map digitization data as well, instead of the "1:2,500 General Purpose Digital Map Data Base Standard."

Digital Cadastaral Map Information
Cadastaral Survey is carried out by prefectural and municipal governments with the promotion by the **National Land Agency.** Currently, only 1/3 of the target area has been completed; the survey and annual progress of the survey is 3,000 sq. km. and more.

In order to make the execution of cadastral survey more effectively, a personal computer system was introduced in 1984. However, the land registration and the land taxing are administered by the Legal Affairs Bureau of the Ministry of Justice and tax offices of a prefectural government, respectively. There is no system linkage among **Digital Cadastral Map Information**, land registration, and taxing so far. Even though surveying procedure, processing softwares, and data format were unified and standardized to some extent, the usage and circulation of the set of data is very limited.

Digital Road Map Data Base Standard (Kamijo, et. al., 1990) Preparation of a digital road map data base permitting computer processing is an indispensable prerequisite for practical application to vehicle navigation. The **Road/Automobile Communication System**, a joint study project towards the overall improvement of road traffic efficiency, was conducted by the **Public Works Research Institute** of the Ministry of Construction together with the research institutions of universities and 25 private companies.

In 1987, the study group made the proposal to develop an up-to-date digital road map database for the entire country. In accordanace with the proposal, the **Japan Digital Road Map Association** was established in 1988. Then the Association has established the **National Digital Road Map Data Base Standard (1st Edition)** in 1988. The preparation of the set of data started in 1988, the basic road network data files have already been completed by March 1989, and full attricute data have been added to the basic road network data by March 1990. Since this database is exclusively used for vehicle navigation, the specification is mainly focused on sophisticated structure of road network and simple structure of background data. This database has already been loaded on car navigation systems of three major Japanese automobile companies of Toyota, Nissan (Datsun), and Mazada.

STANDARD PROCEDURE AND DATA FORMAT FOR DIGITAL MAPPING

Principles of the Standard

"Digital Mapping" used here is a technology to create digital cartographic data through photogrammetric processes and does not include digitization techniques from existing maps nor digital field surveying. It is not only a method to create digital cartographic data but also a photogrammetric surveying and mapping. In this regard, the standardization is focused not only on data format and data code system corresponding to each cartographic entities, but also on the procedure and specification of digital photogrammetric processes. Regarding the photogammetric processes, attention was paid to assure surveying accuracy which was prescribed in certain rules and regulations as for corresponding scale maps. On the other hand, data codes and formats are to be standardized from the viewpoint of data compatibility and data usability in various geographical information systems. The principles of the standardization are as follows:

Purpose of digital mapping

The main purpose of digital mapping is considered to be the creation of digital cartographic data for various geographical information systems. In this regard, digital data set obtained through digital mapping should meet the following requirement.

Assurance of true position data free from displacement or deletion which is inevitable for graphic map.

Data flexibility capable to meet various user objective.

Completeness as the general cartographic information set.

Documents of the standard
Contents of the standard were described in the following seven documents.

Standard Procedure of Digital Mapping
Code system for Digital Cartographic Information
Data format of Standard Exchange File
Manual of Data Acquisition Specification
Recommended Map Drawing Standard
Related File Format and Specification
Quality Control Manual of Digital Mapping

Considering the purpose of digital mapping, standard procedure is determined up to the process of the creation of standard exchange file. Thus, the following processes are determined to be optional.

Creation of Data Management File
Drawing of Graphic Maps
Creation of Structurized Data File

Standard Exchange File
There can be so many files created and used in computer mapping. However, most of them are only temporal or particular to a specific application, which standardization is unnecessary and useless. Here, only the "true position data file" was strictly standardized to guarantee the compatibility. In addition, recommended standards were shown for mapping (drawing) data file and structurized data file.

Code System for Digital Cartographic Information
In order to identify each cartographic entity, ten digit integer composed of two digit "layer code," two digit "data item code," and two digit "area code," and four digit "information code" were prepared. Layer code and data item code are strictly standardized, while area code and information code are optional. Area code is to be used to express locational or topological information. Information code is to be used to sub-classify or re-classify data item code.

Standardized "layer code" are as follows.

Code	Layer
11	boundary
21	road
22	road facility
23	railway
24	railway facility
30	unclassified building
31	low building
32	high building
33	building without wall
34	building belongings
35	building use
41	public utility facility
42	other small objects
51	water front
52	water facility
61	wall, fence, etc.
62	special field
63	vegetation
71	contour
72	sharp
73	control point
75	DEM
81	text

Data Format of Standard Exchange File
Figure 1 shows the generalized data structure. Hierarchical structural with any depth of levels was adopted. There are six record types defined as follows. All records are fixed length of 84 bytes in ASCII/JIS characters.

Index Record
As a header of one complete digital cargographic data set, this recored describes the information of contents and recorded on indepedent file.

Map Marginal Unit Record
The data set is to be divided into the appropriate unit which usually corrsponds to conventional map sheet so that we can void the decrease of usability caused by the huge quantity of data. Each unit is stored in one file. Map marginal unit record contains the information of each unit as name of the area, geodetic coordinates of the unit origin, date of data creation, revision, etc., and is recorded at the top of the file.

Group Header Record
In the hierarachical structure each element corresponding to individual cartographic entity can be grouped. Moreover, elements, group of elements, and group of groups can be grouped. This group header record describes the information on how the group is composed. At level one shown in Figure 1, each group corresponds to each layer.

Grid Header Record
Most of digital mapping data are composed of points and sequence of points (so called vector data). Besides, there can be grided data as digital elevation model data. Grid header record is a reacord to describe the grid origin, interval, size, etc., and is followed by substantial grided data records.

Element Record
This record is the lowest level header record which corresponds to individual cartographic entity which is expressed by one type of substantial data record. This record is followed by substantial data records.

Substantial Data Record
This record describes real geo-coordinates, text, or attributes. There are five types of formats corresponding to three dimensional coordinates, two dimensional coordinates, text, attribute, and grided data.

Principles of Map Drawing
The main purpose of digital mapping is to create digital cartographic data for various geographical information systems. However, drawn or printed maps are still often used in many ways. In this standard, map drawing procedure and specification was not strictly standardized, but recommendations were shown.

Since the standard exchange data files area composed of true position data, it would be rather difficult to read the map directly drawn from the data. In order to improve the legibility of the drawn map, it is necessary to edit or compile the data from the viewpoint of map presentation capability. In this recommendation, most attention was paid on to perform rather automatic editing than manual/interactive editing to improve cost benefit trade off. This direction may cause to reduce map quality.

However, map output is considered to be secondary product in digital mapping, and so, this reduction of map quality is supposed to be allowed.

Structured Data File
The standard exchange data file contains digital cartographic data with the classification code recorded on standard file format. However, for advanced utilization of data as in geographical information system, it is necessary that some data are to be topologically structurized. Structurized data are created for individual geographical information systems and usually unexpected to be exchanged between different systems. Therefore, the structured data file was considered to be out of standardization and only recommendations were shown.

In the recommendation, five standard geometric models are proposed and twelve file formats are proposed for structuring all structural bodies. By combining five standard geometric models, any structural body can be structured.

The standard geometric models are as follows.

Individual point model
Individual line model
Individual plane model
Network model
Area divided model

The twelve files are as follows.

Index file
Structurized file
Point file
Line file
Plane file
Plane definition file
Standard topology file
Extended topology file
External Identifier file
Digital mapping attribute file
User attribute file
Correspondence file to Standard Exchange File

REFERENCES

Akiyama, M., Takayama N., Okuyama S., Hisamatsu F., Kojiroi R., and Kiuchi T., 1988, "Standard Procedure and Data Format for Digital Mapping," International Archives of Photogrammetry and Remote Sensing, Vol. 27, Part B4, pp. 1-7

Geographical Survey Institute, 1987, "Cartographic Work in Japan 1982-1985, "Bulletin of the Geographical Survey Institute, Vol. 32, pp 1-11

Hoshino Y. and Inaba K., 1987, "The Standardization of Digital Mapping," Bulletin of the Geographical Survey Institute, Vol. 32, pp. 12-22

Kamijo S., Okumura K., and Kitamura A., 1989, "Digital Road Map Data Base for Vehicle Navigation and Road Information Systems," Conference Record of Papers presented at the First Vehicle Navigation and Information Systems Conference (VNIS '89) IEEE cat #89CH2789-6, pp. 319-323

Miyazaki Y. and Tusukahara K., 1987, "Digital Map Information in Japan," Bulletin of the Geographical Survey Instaitute, Vol. 32, pp. 23-29

Otake K., Tsurumi E., and Inoue N., 1985, "Collection of Detailed Digital Land Use Data and Its Use," Bulletin of the Geographical Survey Institute, Vol. 29, Part 2, pp. 65-70

Tada H., Morita T., and Nasu M., 1990, "Practiacal Choices for the Building of a Japanese Road Administration Information System," International Archives of Photogrammetry and Remote Sensing, Vol. 28, Part 4, pp. 236-245

ADDRESS LIST OF CONTACT

Geographical Survey Institute (GSI)
(for 3.1, 3.2, 3.3, 3.5, 3.6) Kitasato -1, Tusukuba-shi, Ibaraki-ken

Japan Digital Road Map Association (DRMA) (for 3.8) Hirakawa-cho 1-3-13, Chiyoda-ku, Tokyo

National Land Agency
National Land Information Office
Planning and Coordination Bureau

National Land Survey Division, Land Bureau

Kasumigaseki 1-2-2, Chiyoda-ky, Tokyo

Road Administration Information Center (ROADIC)
Hirakaw-cho 1-2-10, Chiyoda-ku, Tokyo

STANDARDIZATION OF SPATIAL DATA EXCHANGE IN NEW ZEALAND

Patrick Van Berkel
Department of Survey and Land Information
P O Box 170, 103 Thorndon Quay
Wellington, New Zealand

INTRODUCTION

In 1983 the New Zealand national government established Land Information New Zealand (LINZ). This consisted of two groups: the LINZ Board of Management which is the steering group, and its executive body the LINZ Support Group. The purpose of LINZ was and still is to establish and maintain standards pertaining to land information. (LINZ is also responsible for the coordination of LIS/GIS activities, investigating problems associated with matching data from different databases, and demonstrating LIS/GIS concepts.)

Since LINZ has no legislated authority to make use of its standards compulsory it has been a continual effort to get them used. Fortunately many organisations are seeing the wisdom of following standards as they reduce the problems of matching data.

To some extent the establishment of LINZ was ahead of its time as few organisations had digital data. Now there are national, regional and local government organisations who have digital land data, including spatial data. The largest spatial data set is the national digital cadastral database currently being captured by the Department of Survey and Land Information. This database will contain the boundary definition of every property in New Zealand when it is completed in 1992. This data is now being purchased and the problems of data transfer are becoming evident. Major interest in spatial data is also coming from utility organisations (including telecom and electricity supply as they convert their paper records into digital form.

LINZ has close contact with Standards Australia which is also developing spatial data exchange standards and attribute standards.

SCOPE AND GENERAL GOALS

GIS and CAD

It was recognised early on that there is a fundamental distinction in exchanging data between CAD systems as compared with exchanging data between GIS systems. That is a CAD system deals with a digital drawing which contains the attribute data merged in with its graphical representation, whereas a GIS system contains separated spatial and attribute data which may be graphically represented (on a map) in many different ways (depending on scale, selected attributes, etc). Also, in a GIS, the attribute data may be separately analysed and output.

The important characteristics of GIS data which need to be transferred are not colour and line thickness, etc, but the data itself and the relationships. For these reasons CAD standards such as IGES are not considered suitable and a separate GIS standard is needed. (Incidentally it is intriguing to see that the CAD industry is now coming to its senses and new CAD systems are being developed where the attribute and spatial data are being separated!)

Cartographic and Spatial Data

The term "cartographic" implies drawings, and qualities related to output. In most applications, when exchanging spatial data, how it is graphically represented on output is relatively unimportant. It is the data content itself which is of most importance ie, the coordinate data and the attribute data. The symbology (colour, line style, etc) used to output spatial data will depend on what the recipient of the data is trying to convey. This will differ between recipients of the data and will differ in the various outputs generated by one recipient. Thus the terms "spatial" are used, where appropriate, rather than "cartographic" to avoid confusion.

Spatial and Attribute Data

It is recognised that digital land data has two parts: spatial data component and the attribute data. To enable the recipient of transferred data to get the most and quickest use of the data it is necessary that both the spatial data and the attribute data follow standards when being transferred. LINZ is establishing standards in both these areas.

In fact, LINZ is recommending that NZ agencies adopt the LINZ attribute standards in their own databases. This avoids the need to convert between their internal database format and the LINZ format every time a transfer takes place. LINZ has not made any such recommendation on spatial data as it will be many years (if ever) before one spatial data model (eg raster, or polygon/line/point with topology) gains preference over all others.

Goals
Attribute standards have been or are being developed in street address, land use code, national coordinate system, area measurement, territorial local authority code, and Government department code.

A spatial standard is also to be developed. It will be based on the Spatial Data Transfer Standard (SDTS) being developed in the United States. Modifications to SDTS will be similar to those proposed by Standards Australia.

HISTORY AND BACKGROUND

In 1987 a discussion document was written by LINZ and circulated throughout New Zealand explaining spatial data transfer concepts, the need for a spatial data transfer standard, and a review of standards in existence at the time. It distinguished between a standard for the transfer of spatial data and standards for attribute data. A polygon, line, point model, with topology was proposed. Also in the discussion document were proposed formats for the transfer of attribute data such as names, addresses, time, date, etc. Where possible ISO standards were referred to.

A favourable reaction was generally received to the proposal for a standard.

However as time progressed through to 1989 it became apparent that the SDTS would meet our requirements for the transfer of spatial data once a few minor modifications were made to fit the New Zealand situation (such as using the New Zealand Map Grid as the standard coordinate system).

The SDTS has both a spatial file transfer standard and a feature classification standard. In 1989 the LINZ Board of Management adopted, in principle, the spatial file transfer component of the SDTS as the recommended means of transferring files of spatial data. The feature classification was not included at that stage as it was not considered appropriate for New Zealand.

CURRENT WORK

Spatial Data Transfer
New Zealand's major spatial data sets follow a polygon, line, point data model including topology, or are in raster format (eg from photogrammetry or satellite imagery). Thus it is essential that a standard be able to handle these data sets. As explained above the SDTS meets our requirements in these areas. The conceptual basis of the SDTS is part of the standard itself and is further extensively documented in the reports of the U S National Committee for Digital Cartographic Data Standards.

Having approved the SDTS in principle, LINZ awaits the

SDTS becoming finalised and a Federal Information Processing Standard (FIPS) of the United States. It will then be modified to meet New Zealand conditions and approved as the recommended New Zealand standard for transferring spatial data files.

It is anticipated that major GIS vendors will quickly write translators to convert data to and from the SDTS and their internal formats. The availability of these translators will then hasten the adoption of the SDTS by New Zealand agencies.

Entity and Attribute Definitions
The entity and attribute definitions in the SDTS are not included in the NZ spatial data transfer standard. There is no immediate intention to develop a New Zealand standard for entity and attribute definitions. However there are user-oriented working groups looking at entity and attribute definitions one for topographical and city mapping, and another for utility mapping. It is intended to have a first draft of the utility definitions completed this year for circulation throughout the user community.

A partial copy of the U S Entity and Attribute Definitions were obtained in digital form in September 1989 from the U S Digital Cartographic Task Force. These have been loaded into a database for reference.

CONCLUSION

New Zealand is taking an active interest in overseas developments in spatial data transfer standards, and is developing its own standards where overseas standards are inapplicable.

Copies of the LINZ discussion document and the LINZ standards can be obtained from the LINZ Support Group, P O Box 12-271,
Wellington, New Zealand.

REFERENCES

Spatial Data File Transfer Formats

Standards Association Australia, AS2482-1984, "Standard for the Interchange of Feature Coded Digital Mapping Data.

Canadian Council on Surveying and Mapping, 1984, "National Standards for the Exchange of Digital Topographic Data, National Digital Mapping Standards", Topographical Survey Division, Ottawa.

Department of Scientific and Industrial Research, DSIR, 1984, "Computer Storage, Manipulation and Display of Scientific Information on Maps", Department of Scientific and

Initial Graphics Exchange Specification (IGES), Version 2.0, 1983, National Bureau of Standards, US Department of Commerce, Washington DC.

International Standards Organisation, ISO 646, "Information processing ISO 7-bit coded character set for information interchange".

International Standards Organisation, 1976, ISO 3788-1976, "9 track, 12.7 mm (0.5 in) wide magnetic tape for informatin interchange recorded at 63 rpmm (1600 rpi), phase encoded".

International Standards Organisation, 1985, ISO 8211-1985(E), "Information Processing - Specification for a Data Descriptive File for Information Interchange".

Moellering, H., 1984a, ed., "Digital Cartographic Data Standards: Examining the Alternatives, Report No 4", Issues in Digital Cartographic Data Standards, National Committee for Digital Cartographic Data Standards (NCDCDS), Columbus, Ohio, USA.

Moellering, H., 1984b, ed., "Digital Cartographic Data Standards: Examining the Alternatives, Report No 5: A Working Bibliography for Digital Cartographic Data Standards", NCDCDS, Columbus, Ohio, USA.

Moellering, H., 1985, ed. "Digital Cartographic Data Standards: Examining the Alternatives, Report No 6: An Interim Proposed Standard", NCDCDS, Columbus, Ohio, USA.

Moellering, H, 1987, ed. "Digital Cartographic Data Standards: Report No 8: A Draft Proposed Standard for Digital Cartographic Data", NCDCDS, Columbus, Ohio, USA.

New South Wales (NSW), 1986, "New South Wales Standard for the Interchange of Digital Land Information Data, Volume 1: Technical Specifications and Tables, and Volume 2: Description of Tables with Examples", NSW State Land Information Council, Land Titles Office Building, Sydney.

Peuquet, D.J., 1984, "A Conceptual Framework and Comparison of Spatial Data Models", Cartographica, 21(4), 66-113.

US Geological Survey (USGS), 1983, "USGS Digital Cartographic Data Standards" USGS Circular B95, National Mapping Division, US Geological Survey, Reston, Virginia.

Van Roessel, J.W. and E.A. Fosnight, 1985, "A relational approach to vector data structure conversion", Proc Auto Carto 7, Washington DC.

White, M.S. Jr, 1984, "Technical Requirements and Standards

for a Multipurpose Geographic Data System, The American Cartographer, 11(1), 15-26.

Williams, R.D., 1985, "POLAR Users Manual", Internal Report WS905, Hydrology Centre, Ministry of Works and Development, Christchurch.

Working Party to Produce National Standards for the Transfer of Digital Map Data, 1986, "The National Transfer Format, Final Draft", Ordnance Survey, Southampton, England.

Attribute Data Transfer Formats

ISO 2014-1976: Writing of calendar dates in all-numeric form. International Standards Organisation.

ISO 2711-1973: Information processing interchange Representation of ordinal dates. International Standards Organisation.

ISO 3307-1975: Information Interchange Representations of time of the day. International Standards Organisation.

ISO 6523-1984: Data Interchange Structure for the identification of organisations. International Standards Organisation.

ISO 6709-1983: Standard representation of latitude, longitude and altitude for geographic point locations. International Standards Organisation.

LINZ Support Group, 1986, LINZ Standard for Departmental and District Office Codes.

LINZ Support Group, 1985, LINZ Standard for Geographic Reference.

LINZ Support Group, 1986, LINZ Standard for Land Appellation.

LINZ Support Group, 1986, LINZ Standard for Territorial Local Authority Name.

LINZ Support Group, 1987, LINZ Standard for Area Measurement.

LISAC, 1982, Standard Data Formats, Western Australia Land Information System Advisory Committee.

Standard Land Use Code Committee, 1984, "Report of the Standard Land Use Code Committee (incorporating a draft NZ Standard Land Use Classification)", Department of Statistics.

STANDARDIZATION EFFORTS IN NORWAY

Olaf M. Østensen
Project Manager
National Geographic Information Centre
The Norwegian Mapping Authority
N-3500 Hønefoss, Norway
tel. + 47 67 24 100
fax + 47 67 26 190

ABSTRACT

The standardization work in Norway regarding digital spatial information is described. Both the present status and future plans are covered. The present use and experience is also mentioned. Norwegian ambitions are high both with respect to transfer formats, electronic data interchange and data modelling. The different standardization levels are characterized and put into a total context.

INTRODUCTION

There is currently considerable growth in the use of GIS in Norway. This is a recent development that has taken place the last couple of years. It is a consequence of the acceptance of GIS among not only the map producers, but also the users of geographic information. The governmental program for information technology has recognized geographic information as one of the fields with the greatest potential for growth with regard to new technology. The Royal Norwegian Council for Scientific and Industrial Research have supported the industry and user organizations through a program for Geographic Information Technology (GIT) for some years now. But although we see a considerable growth, the situation is still characterized by a lack of competence and also a lack of important features in the available systems, causing many potential users still to "sit on the fence".

One important factor to be aware of is that there is a growing consciousness about common administration of information shared among many users. There are several projects, both at national and regional level, currently dealing with this aspect. All these projects rely on a standardized way of transferring information between a repository and local user applications. The most important case is the national service described below in the last section.

Another factor is that there is a very positive attitude towards standardization. Most users see this as a protection of the investments they make in the establishment of basic information.

This presentation shows that the scope of Norwegian activities is perhaps broader than that in several other countries at present - this designates the Norwegian ambitions, and the various parts described below must be considered as integral parts of an total approach to the standardization issues.

The the following features are described:

- o the 'traditional' standard transfer format (essentially a file format)
- o EDI (electronic data interchange) features
- o a standard conceptual data model for common data (essentially features conforming to the content of our traditional map series)

Further information can be obtained from:

The Norwegian Mapping Authority
SOSI secretariate
N-3500 Hønefoss, Norway
tel. + 47 67 24 100
fax + 47 67 26 190

HISTORY AND BACKGROUND

The work leading to our present standard was started in the late 70's. The Ministry of Environment then defined a project "Samordnet Opplegg for Stedfestet Informasjon" - i.e., "Coordinated Approach to Spatial Information". The acronym for this project was, as you see, SOSI - a name that since then has been attached to this work. The main purpose of the project was to propose a system for coordination and joint management of the work with spatial information within a decentralized network of data producers and data users.

One of the reports (1980) that came out of the project suggested the definition of an exchange format, and also suggested the syntax of such a format. This report was based on studies of Canadian and German cases. It concluded with the proposal of utilizing results from similar although unrelated projects within large Norwegian industrial companies. That proposal was not realized, but the ideas behind the syntax became the starting point of the development.

Prior to the realization of a national standard, and even still to some minor extent, a series of very simple formats have been quite popular. In general these formats can be described as consisting of sets of 'objects', each having a thematic code together with graphic representation, e.g., a series of planar coordinates. Of course these formats are quite insufficient to express structures or complex attributes of any kind.

The responsibility for the standardization in the period 1972-1985 was with the Ministry of Environment. From 1981 the process of defining the standard format slowly gained momentum. A questionnaire and a seminar was held in the autumn of 1982. The format was first formally defined and published in 1985, and a complete version with object definitions and attribute coding appeared in 1987. In spite of the long way from con-ceptual ideas to a published format, versions of the format were in use within the governmental mapping institutions as early as 1983.

In the meantime, the three governmental map institutions, the Norwegian Geographical Survey, the Hydrographic Service and the County Mapping Offices were combined to form the Norwegian Mapping Authority in 1986, and responsibility for standardiza-tion work was transferred to the new institution.

Up to now the Norwegian Standards Association has not been involved in this work. On the other hand we have closely cooperated with the Norwegian Association for Cartography, Geodesy, Hydrography and Photogrammetry and the Norwegian Association of Surveying Companies.

In the Mapping Authority the work has recently been organized within the large project called the National Geographic Information Centre (NGIS, in Norwegian). This centre is responsible for the building of our large, central database of public geographic information and the services supported by the centre as well as the information resource management required for this purpose, including data structuring and standardization issues. Next year (1991) the project will be transformed into a part of our permanent organization.

Norway is also contributing to the international standardization efforts, most important is the Norwegian Hydrographic Services work with the International Hydrographic Organization (IHO) on a standard for the exchange of electronic charts.

THE CONTEXT OF STANDARDIZATION

We will make some observations and develop basic concepts.

Any database - or other file organization - can be seen to consist of two parts: i) a data model, and ii) physical data in accordance with the model. The data model represents the semantics of information. The purpose of an information transfer is then, ideally, to transfer both the physical data and their meaning or conceptual model.

Unless the source database and the receiving database have the same conceptual model of the data involved in the transfer, the transfer is in principle imperfect - that is, something is lost in the transfer. In most cases such imperfect transfers can be accepted, and the receiving system will try to interpret the transferred information as well as possible.

This interpretation can be based on several mechanisms, basically of two different types: i) the transfer includes a complete description of the source data model, or ii) the transfer format implements some sort of a generic data model which then both source and receiver have to interpret as well as possible.

We also see hybrid solutions, where for instance attribute information descriptions are transferred along with data while geometry is based on a common underlying model.

The present Norwegian standard, or - more precisely - practice of the standard, conforms to this latter description.

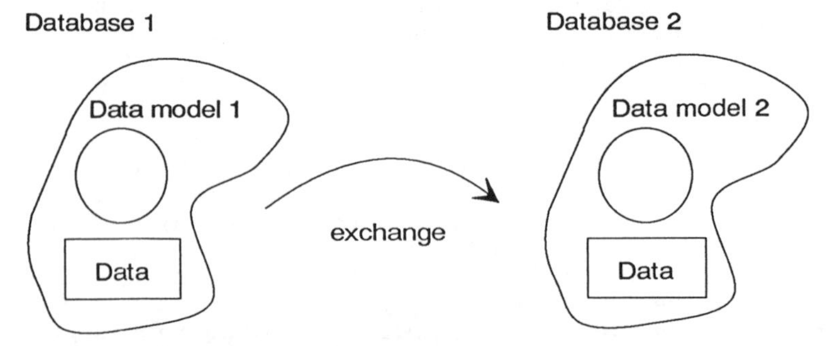

Figure 1. Principle of information interchange

THE TRANSFER FORMAT - SOSI

Introduction

The SOSI format is based on the following principles:

- o flexible, dynamic format
- o description of data along with data itself
- o allows data compression

It is, of course, hardware and software independent and is today implemented on a range from the largest mainframe to PCs.

SOSI does not really describe a conceptual data model, but rather a minimum way of representing basic geographic (or geo-metric) building elements, i.e., object types built up of points, lines/curves and polygons (or, more general, two-dimensional objects).

The SOSI standard consists of three parts - a syntax, a semantics and sets of coding values for feature and other attribute coding.

Four levels of complexity are defined:

- o Level 1: More or less unstructured 'spaghetti'-data and very simple classification and attribute information.
- o Level 2: As above, but more attribute and coding information, information connected to points in geometric constructs.
- o Level 3: As above, including also point connectors or nodes. Covering network structures.
- o Level 4: As above, including also complete geographic objects (areas, curves etc.). Serial numbering and referencing.

Complete documentation is given in the latest SOSI-manual, version 1.4.

Structure
The structure is composed of:

- o a header
- o user definitions
- o user data
- o a trailer

Header
The header is a flexible part describing general information about the user data, e.g., who is the owner, who is the producer, coordinate system, general accuracy, currentness. Definitions in the header are considered defaults for the user data, but will be overridden by classification within the user data itself.

User definitions
The standardization committee maintains a set of standard definitions which is considered implicit in any occurrence of

a SOSI-file. Any user can enhance this standard set by describing his own definitions of structures.

User data
User data is organized in a hierarchical structure of groups of information and basic information. Basic information elements are composed of a descriptive name and a value:

<Basic element>::=<Element name><Value>

Group elements are constructed from basic elements and also given a group name. Groups can contain any level of subgroups, thus describing a tree structure of arbitrary height.

Every basic element and group element has to be defined, either implicitly as a part of the standard or explicitly as part of the user definitions.

In an instance of a group element, one can compactify by just giving the group name and then the values of the group tree parsed from left to right. The parsing rule gives a one-to-one mapping of values to implicitly given description in the form of basic element names. One can also enhance a group by adding new groups to the defined tree, but then this must be done without creating inconsistency between values and descriptions.

Any group can be given a serial number, which can be referred to in other parts of the user data.

All in all, the result is a powerful and flexible format.

The trailer
The trailer is just a specific name .SLUTT (== .END) designating the end of one set of a SOSI-unit. Other units can follow within the same physical file or transmission.

STATUS

The SOSI format has been defined for the following areas:

- o general map information, scale 1:5,000 and smaller
- o technical mapping, scale 1:200 - 1:1,000 (local authority responsibility)
- o road mapping, cooperation with the Public Roads Administration
- o local planning, i.e., municipal planning

The format is in everyday use both in deliveries from the Mapping Authority, within large governmental institutions and at local authority level.

The Public Roads Administration and the Telecommunication Administration - two of the larger public enterprises and users of GIS - require the use of SOSI in their contracts with private surveying companies.

All major vendors operating in Norway have developed conversion routines to and from their applications, including systems like Arc/Info, Intergraph, AutoCAD and SysScan.

EXPERIENCE AND PRACTICAL USE

Experience up to now is that the format functions very well and is specially suited for complex data structures where the preceding formats fall short. The negative part is that a lot of systems in use are unable to represent the more advanced features allowable within the format - e.g., multiple identification and feature codes. In spite of this, everyone seems to acknowledge the format as the only possible way of achieving standardization.

Parallel to the development of the standard format, the Norwegian Mapping Authority has developed a computer system for digitizing (data capture) and editing completely based on the format as underlying data structure. This has been a key factor in the success of the practical use of the format. This practical use also constitutes the testing of the format. The format has proved to function in everyday, real life use - perhaps the most serious testing possible.

This practical use means that information corresponding to several thousands of map sheets with different complexi-ties and scales have been stored in the format. In this sense the format has served an important role as a system- independent long-term storage format. Of course this is a possible side effect of any transfer format. In a time period where data base systems change rapidly, or has not been decided, this functionality is very important.

The format has proved to be suitable both for physical media transfer like magnetic tapes, diskettes and optical media including CD-ROM (an experimental CD-ROM was produced in 1987), and for communication lines from LANs (Local Area Networks) to WANs (Wide Area Networks). This is not surprising as it is in a readable ASCII-format. Provisions have been made to account for different national character sets or different vendor-dependent character sets (e.g., PC-8, 7-bit Norwegian character set, EBCDIC and so forth).

The mechanisms provided for data compression have proved to give a reasonably compact format competetive with fixed format binary formats. The compression schemes retain readability as well as reduce default information, empty fields and repeating information to a minimum.

ONGOING ENHANCEMENTS AND FUTURE PLANS

The format is under continuous development. We have seen a rapid acceptance over the last couple of years, and also a demand for handling new fields, both of information types and of professions.

Up to now the format has only handled vector type information (and alphanumeric, non-geometric data). We are now starting to define

- o raster information
- o pictorial information

By raster information we here mean pixel information, both monochrome and colour, with an underlying map coordinate system. By pictorial information we mean more general pixel or vector information without a map geometry, for instance a scanned colour photograph of an interesting object, architectural drawings of a building or a CAD-drawing. These non-mapping graphical objects will be described in general standard formats or de-facto standard formats.

Among formats to be considered are CCITT (Comité Consultatif de Télégraphic et Téléphonie) gr. 4 facsimilieformat and ODA/ODIF (ISO 8613).

The following professional areas are currently being worked on:

- o utilities, telecommunication, energy, water
- o natural resources and environmental data

A NEW LEVEL OF STANDARDIZATION - FGIS

The FGIS (Felles GIS - Common GIS, or, better, Common Geographic Information Standards) project was initiated in the summer 1989 and is finished this year.

The scope is to establish

- o a common conceptual data model
- o a geographic data definition language
- o a geographic database API (application program interface)
- o a standard user interface for review, catalogue information and ordering
- o a geographic EDI management system

	Present	Current work	Future
Data base level		FGIS	FGIS
Data content level	prop. syntax SOSI	prop. syntax SOSI	prop. syntax SOSI ISO 8211
Message level	NGIS proprietary	EDIFACT	EDIFACT
Communication level	SNA	SNA,TCP/IP	X.400/X.500,FTAM
Link level	IEEE802.2,.3,.5 X.25 WAN	IEEE802.2,.3,.5 X.25 WAN	IEEE802.2,.3,.5 X.25 WAN

Figure 2. Overview of status and plans

- o a prototype of such a system based on ANSI/SPARC application model and a relational database

It is very important to understand that we are not defining another GIS, but a standard for geographic information resource management.

Today we have the specifications and a simple prototype model illustrating the concept. We are in the process of industrializing the results of the project. The result of this process will hopefully be a company delivering products based on the specifications on an open platform, that is, on a variety of hardware and software technologies from different vendors.

It is a prerequisite that the FGIS solutions shall be able to coexist along with today's commercial GIS like Arc/Info, Intergraph, SysScan, etc. More specific, the main idea is that FGIS shall be basis for host systems for a variety of GIS solutions, enabling a public or private enterprise to keep their information in an optimal repository and be able to use a number of different GIS tools optimized for special tasks.

The FGIS data definition language will be based on a geographically enhanced ER-model (entity-relationship - model).

We have not yet agreed upon a standard data model of common geographic information. This is the task of a working group starting the autumn of 1990. A rather exhaustive example has been worked out jointly within the FGIS project and a regional GIS project in Mid-Norway. This will be a reasonable starting point for a standards proposal.

The FGIS Kernel is an application program interface (API) to the database management system giving an object-oriented way of seeing the database. In the prototype the kernel is developed using a 4GL called Uniface which is based on the ANSI/SPARC model. The DBMS is a relational DB, but can be substituted with any kind of database, e.g., an object-oriented database when commercially and technically feasible.

The FGIS EDI Management System is based on the ideas of the previous section, and thus fully in compliance with the Norwegian OSI-profile, NOSIP.

	Services	Plus	Minus
Standard data model	As below	No conflicts in concepts	
EDI functions	Electronic interchange - message based - catalog ordering - data ordering - updating messages	Fast, flexible, automatic features, rich set of services	Possibly conflicts in data model
Transfer format	Media transfer - magnetic tape - diskette - optical media - point-to-point electronic	Simple	Slow, rely on manual handling

Figure 3. Different elements of standardization

There are good reasons to believe that FGIS products will gain a large market share in Norway and thus establish a de facto standard for geographic information resource management. Especially the local authority market is important in this respect as more than 50% of geographically based information originates in the municipalities. We know from experience that local authorities have a very positive attitude toward standardization. Today many of them are uncertain about the choice of GIS solution, mainly because they are very concerned about information resource management in a heterogeneuos user environment where today's systems often have very poor solutions. We think products following the FGIS specifications will fill in this gap.

THE NATIONAL SERVICE NGIS - BASED ON STANDARDS

An important driving force in the recent development of the SOSI format and the other standardization elements has been the establishment of a national service called the National Geographic Information Centre, abbreviated NGIS in Norwegian.

This is an open service where anyone can subscribe to access the database. The service is based upon many of the features previously described in this article.

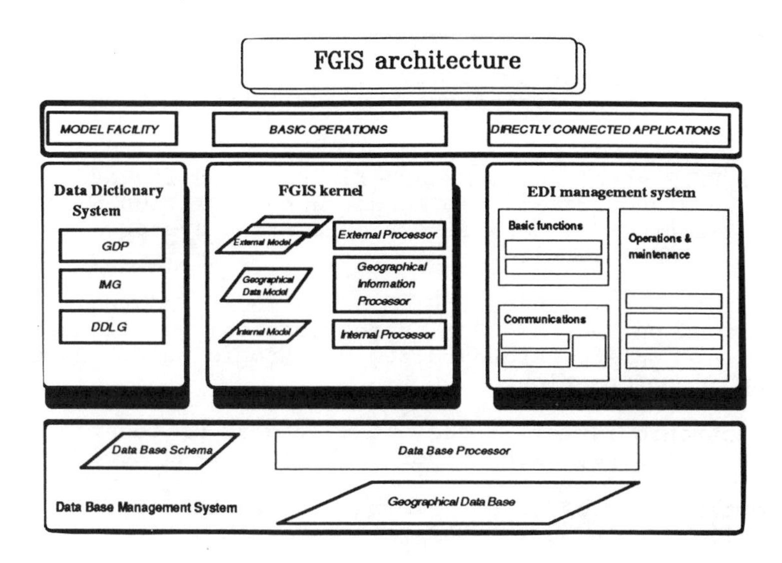

Figure 4. The FGIS architecture

NGIS is completely on-line based, but will not give the customers an interactive GIS tool. Rather it is an EDI-application dealing with geographic information. The first products available are the GAB-database covering all real estate, addresses and buildings (after 1983) in Norway (attribute information), the graphical representation of the real estates, road database for public and private roads and the coast line of Norway. Other information types will be included in a continuous process. Both raster and vector information will be available.

The project was started in 1988 and the service is available this year, but will attain commercial value with a substantial information content from 1991/01/01.

As a very important feature NGIS will provide integrated information in the sense that data types which are naturally related in the real world will also be related in the data base. Up to now this has not been possible, neither in the analog material nor in the digital production datasets or -bases. This of course means new ways of establishing information and new routines for maintaining the information.

But it also made it necessary to include the first kind of message handling to the SOSI format, i.e., the handling of update messages:

- o new objects
- o changed objects
- o deleted objects

NGIS has also made it necessary to define intermediate solutions to the data model issue and for communication services.

When the FGIS specifications are finished and we have completed the standardization process, a new solution based upon that platform will be the next generation system underlying the NGIS service.

Then, with the new generation NGIS and local and regional solutions based on the same FGIS specification, we will have established a complete, standardized electronic marketplace for geographic information.

This electronic market is the ultimate goal of today's efforts and will give, among others, the following benefits:

- o an optimal use of joint information and thereby minimize double work
- o a very flexible and fast access to information
- o a better overview over existing information, for instance by issuing a broadcast of the following type:

  Who has xx information in xy area?
- o a complete standardized conceptual model of common data
- o a distributed and decentralized service covering all authority levels

REFERENCES

ISO 8211, 1985, Information Processing - Specification for a data descriptive file for information interchange.

ISO 8613, 1989, Information Processing - Text and Office Systems; Office Document Architecture and Interchange Format.

ISO 9735, 1988, Electronic data interchange for administration, commerce and transport (EDIFACT) - Application level syntax rules.

Ministry of Environment, 1985, SOSI - et standardformat for digitale geodata (SOSI - a standard format for digital spatial data).

Norwegian Computing Centre, 1977, Samordnet Opplegg for Stedfestet Informasjon (Coordinated Approach to Spatial Information).

Norwegian Computing Centre, 1978, SOSI II, statusrapport okt. 1978 (SOSI II, status report, oct. 1978).

Norwegian Mapping Authority, 1990, SOSI formatet, versjon 1.4 (The SOSI format, version 1.4).

Norwegian Association of Surveying Companies, 1990, Retningslinjer for produksjon og leveranse av stedfestet informasjon (Guide-lines for production and delivery of spatial information).

Norwegian Mapping Authority, 1990, FGIS - Konseptbeskrivelse (FGIS - Conceptual description).

Statskonsult, 1990, Forslag til Norsk OSI-Profil - NOSIP (Proposal for the Norwegian OSI Profile - NOSIP).

THE SOUTH AFRICAN STANDARD FOR THE EXCHANGE OF DIGITAL GEO-REFERENCED INFORMATION

Antony K. Cooper
Centre for Advanced Computing and Decision Support,
CSIR, P O Box 395, PRETORIA, 0001, South Africa.

and

Derek G. Clarke
National Land Information Services,
Chief Directorate: Surveys & Mapping,
Private Bag, MOWBRAY, 7705, South Africa.

ABSTRACT

This chapter describes the development of the South African standard for the exchange of digital geo-referenced information, from the establishment of the project team that drafted the first version of the standard, through to the present and the work of the maintenance committee for the standard. A summary of the concepts used in the standard and an outline of the standard are also given. The chapter concludes with a comprehensive list of references to published work related to the standard.

INTRODUCTION

The South African standard for the exchange of digital geo-referenced information (Clarke, et.al., 1987b) was published during September 1987. Since then, it has been formally accepted by the Co-ordinating Committee for the National Land Information System as the standard for South Africa, and as such, will be supported by all government departments. Usage of the standard has been small as digital geo-referenced information is only starting to become available. An updated version of the standard is in preparation. It will consist of two parts, namely the standard itself and a user manual.

The standard was informally known as SWISK 45 (which was

the CSIR's report number for the version published in 1987), but it is now known as the **National Exchange Standard**, or by its abbreviation, **NES**.

THE INITIAL DEVELOPMENT OF NES

In April 1986, a project team was established formally to draw up proposals for a South African national standard for the exchange of digital geographically referenced (geo-referenced) information. The project was funded by the Foundation for Research Development of the CSIR, through their National Programme for Remote Sensing. It was a joint project between the National Research Institute for Mathematical Sciences (NRIMS - now incorporated into the Centre for Advanced Computing and Decision Support) of the CSIR and the Chief Directorate of Surveys and Mapping (CDSM). The team was led by Derek Clarke (CDSM) and Hester van Rooyen (NRIMS), with the other members being Prof Elri Liebenberg (University of South Africa) and Antony Cooper (NRIMS).

Informally, the project team began working at the end of November 1985, when a workshop was held to plan the strategy of the team and determine its goals. Present at the workshop were a few NRIMS staff members, as well as Prof Ray Boyle (University of Saskatchewan, Canada).

When the project began, no reasonable standard for the exchange of digital geo-referenced information existed. However, a number of exchange standards were completed within the six months either side of the completion of NES. The project team began by studying the work done on these exchange standards in other countries - in particular, in Australia (SAA, 1981), the United States of America (Moellering, 1985), Canada, the United Kingdom (Haywood, 1986) and by the International Hydrographic Organization (CHS, 1985; NAWG 1985). In mid-1986, to gauge the state of the use of geographical information systems (GIS) in South Africa, and the need for an exchange standard, the project team distributed a questionnaire (Clarke, et.al., 1986a) and a glossary of GIS terms (Clarke, et.al., 1986b) to those users and potential users that the project team were able to identify in Southern Africa. The replies received indicated that few organizations in South Africa had much experience with GIS or automated cartography and that there was a need for an exchange standard. These mailings were followed up by visits to a number of these organizations for discussions on the nature and content of the exchange standard. In addition, a number of workshops were held around the country.

In March 1987 the draft exchange standard (Clarke, et.al., 1987a) was completed and distributed for comment. Unfortunately, the response was very poor - written comments were received from only two organizations. Nevertheless, the project team were able to improve the draft version. This final version (Clarke, et.al., 1987b) was released in September 1987. The project team submitted their final report to the National Committee for Remote Sensing (Clarke, et.al., 1987d), and then disbanded.

In December 1987 the Co-ordinating Committee for the National Land Information System, at the request of the National Committee for Remote Sensing, accepted the responsibility for the National Exchange Standard.

SCOPE OF NES

The National Exchange Standard (NES), was designed to cater for all forms of digital geo-referenced information. It was felt that all the different types of geo-referenced information (such as cartographic, hydrographic, topographic, cadastral and thematic information) shared the same fundamental structure, and thus could be exchanged using the same exchange standard. As such, NES caters for vector, raster and alphanumeric information, as well as topological information (information on the spatial relationships inherent in the data), information on the quality of the data being exchanged and alternate spatial attributes (multiple versions of the digital representation of an entity).

The current version of NES caters for two and three dimensional spatial information.

NES uses a relational structure to exchange the digital geo-referenced information. This makes the standard flexible, as additional relations may be added as and when needed, without invalidating data sets prepared using an older version of the standard. It also makes it easier to use as the user need only use those relations that concern him. Global information, such as offsets for coordinates and the projection system and reference surface used, preceeds the geo-referenced information in the data set being exchanged. Information on data quality is included within the geo-referenced information as free text.

NES is dependent on computer industry standards for the definition of physical media used for the exchange of digital geo-referenced information.

THE CURRENT STATUS OF NES

Maintenance Committee

The Co-ordinating Committee for the National Land Information System established a sub-committee, the Standards Committee, which is responsible for all standards related to the National Land Information System. At this stage, the two standards produced by the committee are NES and a standard for data. With reference to NES, the committee's functions are:

- To promote the use of the national standard.
- To provide an advisory service to interested

organizations in matters relating to the exchange of geo-referenced information.

- To promote research in and investigate matters relating to the exchange of geo-referenced information and to keep abreast of developments in the computer and information industries and similar exchange standards in other countries.

- To establish and maintain a directory of organizations with geo-referenced information and their conditions for exchanging this information.

- To provide an information service on the availability of geo-referenced information.

- To maintain the national exchange standard by receiving and co-ordinating users' requests and comments and then issuing revisions of the national standard when deemed necessary.

- To prepare and maintain a user manual.

The maintenance committee has prepared a questionnaire (NESC, 1989) on the availability of digital geo-referenced information in South Africa. The questionnaire asks the user to specify what information is available, specifying the feature classes, geographical extent and non-spatial attributes, and for some data on the quality of the information. There are also a few questions on the conditions of supply.

The published version of NES (Clarke, et.al., 1987b) was intended to provide a precise definition of the standard, albeit with a few introductory chapters. A user manual is therefore needed. This has been borne out by those who have used the standard (Greenwood, 1988). The user manual will provide a step-by-step guide on how to develop and use interface programs for the exchange standard, using examples.

There are a number of errors in the published version of NES (Clarke, et.al., 1987b). While none are errors of logic (they are mainly typographical) they could cause problems for people reading the document. In addition, the exchange standard has to be improved in a number of areas. The classification scheme is incomplete. The exchange of information on quality must be improved, both by quantifying the information and by providing guidelines on what must be provided. Improvements can also be made to the relations that cater for non-spatial attributes and for raster data. This version of the standard is currently under revision.

Concepts Used In NES

In this section, the basic concepts used in NES will be introduced briefly, and NES will be described. These definitions are taken largely from (Cooper, 1989a) and are fairly brief - for more detail, the reader is referred to the published version of NES itself (Clarke, et.al., 1987b).

Features, Attributes & Classification: In NES, **features** are the basic entities of digital geo-referenced information. A simple feature is a set of one or more uniquely identifiable objects in the real world where the defined characteristics of the objects are consistent throughout all the objects. Features can be man-made or natural, real or abstract. These defined characteristics are known as the **attributes** of the features, and can be **spatial** (that is, dependent on the feature's position in the n-dimensional space) or **non-spatial** (that is, independent of the feature's position - also known as the descriptive information of the feature).

Classification is the arrangement of features into classes or groups and should be done on the basis of the **qualitative** characteristics of the objects, such as their function, and not on their **quantitative** characteristics. A feature's classification should be based on those of its characteristics that are least likely to change. There is a fine distinction between the non-spatial attributes of a feature and its classification because for different users, different criteria for classifying the information apply. One could even consider the classification itself to be a non-spatial attribute (Cooper, 1987a). While NES may be used with any classification scheme, the standard includes a skeleton classification scheme based on a variable-level hierarchical model for classification (Clarke, et.al., 1987b; Cooper and Scheepers, 1989).

Spatial attributes. A **spatial attribute** is an attribute whose value is a subset of any n-dimensional space - this version of NES caters for only two and three dimensions as they are the most commonly used. Should further dimensions become widely used, NES will be expanded to cater for them, which should not prove difficult. Note that in the current version, temporal values may be recorded as non-spatial attributes. Spatial attributes may be **vector** (that is, positional data recorded as coordinate tuples forming nodes, chains, etc) or **raster** (that is, data expressed as a tesselation of cells, with spatial position implicit in the ordering of the cells).

NES provides four fundamental types of two-dimensional vector spatial attributes, namely **nodes**, **chains**, **arcs** and **regions**, and the fundamental raster spatial attribute, the **matrix**.

A **node** is a 0-dimensional object with an n-tuple of coordinates specifying its position in n-dimensional space. The position of a **point feature** is described by a single node.

A **chain** is an ordered undirected sequence of n-tuples of coordinates with a node at each end. An **arc** is any continuous part of the circumference of a circle with a node at each end. The position of a **line feature** is described by a set of one or more chains and/or arcs, which do not necessarily form a continuous object.

A **region** is the interior of a continuous and closed sequence of one or more chains and/or arcs, known as the region's outer boundary. The position of an **area feature** is described by a set of one or more regions, which do not necessarily form a continuous object.

A **matrix** consists of an n-tuple of coordinates, specifying its origin, and an m-dimensional rectangular tesselation of data values encoded in a pre-defined format. The position of a **grid feature** is described by a set of one or more matrices, which do not necessarily form a continuous object.

Compound features are those which consist of one or more other features. This allows the user to build a hierarchy of features, for those occasions when the individual constituent features have their own non-spatial attributes (and classification), but together they have other additional non-spatial attributes and a classification.

Topology: NES caters for two topological relationships, namely **coincidence** and **exclusion**. Coincidence refers to the sharing of common sets of coordinate tuples, and is modelled by having more than one feature share the same spatial attributes. Exclusion refers to area features that consist of regions that wholly contain other regions that do not form a part of the area feature. Exclusion is catered for explicitly through two relations in the exchange standard.

The concepts of features, spatial and non-spatial attributes, classification and topology are generally shared by NES with other exchange standards, though the terminology might differ. A concept unique to NES is that of alternate spatial attributes.

Alternate spatial attributes: A feature has **alternate spatial attributes** when it is represented by a number of different sets of spatial attributes, where each set defines fully the location of the feature. An **alternate spatial attribute scheme** determines the manner in which the different alternate spatial attributes are related to their features. There are two main reasons as to why a feature would have alternate spatial attributes.

Firstly, in an area with a high density of features, the graphical representation of the area (be it on a computer screen or hard copy) would be messy, unless the display of some of the features could be suppressed, or unless some of them could be represented in a simplified manner. However, for analysis on the spatial attributes of the features, one would prefer to retain the spatial attributes of all the features, and to retain them to as much detail as possible. Alternate spatial attributes allow one to keep different versions of the spatial attributes for a feature to solve this problem - at one level, the alternate spatial attributes are for display, while at another level they are for analysis.

Secondly, if one deals with data at greatly disparate scales, one would like to retain different, scale dependent, versions of the spatial attributes of those features which appear at both small and large scales - automatic generalization of spatial data from a large scale to a small scale is still an interesting research area, and it is not possible to create large scale spatial data from small scale data! Again, alternate spatial attributes allow one to keep more than one set of spatial attributes for a feature.

In NES itself, there is an entry in the **Global Information Section** (see below) which determines whether alternate spatial attributes are used in the data set being exchanged, and if so, which scheme is used. If they are used, then in the **Geo-referenced Information Relations**, the field **Alternate Spatial Attribute** is used in every relation between features and spatial attributes, as well as in the two relations which define the type of the feature (point, line, etc) and its planimetric spatial domain. If alternate spatial attributes are not used, then the field is ignored completely.

There is also a relation in the Geo-referenced Information Relations for exchanging, with the data set, an alternate spatial attribute scheme - no such scheme is defined in the current version of the exchange standard.

Information on the quality of the digital data: The American National Committee for Digital Cartographic Data Standards (NCDCDS) identified the nature of information on the quality of digital geo-referenced information, and which information should be recorded (DCDSTF, 1988).

Although some exchange standards allow for the encoding of some forms of information on the quality the digital data, NES has followed the lead of the NCDCDS and allows the information on quality to be exchanged as free text only. A relation is used which may be included as often as necessary in amongst the Geo-referenced Information Relations. The granularity of the information on quality can thus vary from coarse (referring to the whole data set) to fine (referring to a section containing only one instance of a particular relation) (Cooper, 1987a).

Only once the quantification of information on the quality of digital geo-referenced information is well understood, will the exchange standard address the encoding of such information on the quality of the digital data.

The relational model of the exchange standard: A data set

encoded in the format defined by an exchange standard is not a data base - it is merely a set of data that has been extracted from one data base with the purpose of being incorporated into another data base. The authors have noticed that some users confuse these data sets being exchanged, with data bases.

To be successful, an exchange standard must be independent of the data bases that might be interfaced to it. There are three common models for data structures, namely the hierarchical, the network and the relational. NES uses a relational model because it is inherently modular and more flexible than the hierarchical or network models. In a relational structure, the data are represented in a single uniform manner, and thus operations on the data are robust and simple to implement. When creating a data set in the format defined by NES, the user merely omits those relations for which he has no data.

It is easy to add new relations to NES - in fact, data that can be exchanged through the relational structure of the current version of NES should always be able to be exchanged through future versions of NES, no matter how many new relations are added to cater for new concepts or types of data. This is achieved by adding new relations and leaving the existing ones as they are, rather than modifying the existing relations.

It is desirable to have a degree of normalization in data in a relational form (Van Roessel, 1987). There are some relations in NES for which normalization was not feasible due to the excessive storage and processing overheads that would be introduced. For example, the records in the relation containing the internal coordinates of chains have variable numbers of fields (one field for each coordinate). For the rest of the relations, an attempt was made to normalize the relations to the third normal form. This required the introduction of a **Sequence Number** field to the keys of those relations where the keys were not unique, for example the relation relating classification to feature - any feature class may have many features with that classification. However, the sequence number appears only in the document describing the standard and not in the data being exchanged. As the data in the data set have an inherent ordering, the sequence number is implied by the record's position in the data set.

As an example, the following are the relations which relate an area feature to its classification and its spatial attributes:

- Feature/classification which relates:
  Feature ID <==> Classification

- Feature/feature type which relates:
  Feature ID <==> Feature type

- o Area feature/included regions which relates:
  Feature ID <==> Region ID

- o Region/chains & arcs & direction which relates:
  Region ID <==> Indication of chain or arc U Chain ID U Direction indicator

- o Chain/nodes & coordinate tuples which relates:
  Chain ID <==> Node ID U Node ID U Length of chain U Data ID

- o Node/coordinate tuple which relates:
  Node ID <==> Coordinate tuple

- o Chain data which relates:
  Data ID <==> Coordinate tuples

Relation 1 classifies the feature, relation 2 identifies the feature as an area feature, relation 3 connects the area feature to its region spatial attribute, relation 4 performs the topological link between the region and the chains and arcs which form its boundary (specifying whether the chains and arcs are used forwards or backwards), relation 5 links the chains to their start and end nodes and to their internal coordinate tuples, relation 6 specifies the coordinate tuples identifying the locations of the nodes and relation 7 contains all the internal coordinate tuples for the chains.

Outline of NES

A data set encoded in the format defined by NES consists of three parts, namely the File Identification, the Global Information Section and the Geo-referenced Information Relations, in two separate files. The File Identification is in one file, providing a quick, easy to read description of the data set, and the other two parts are in the other file. The Global Information Section provides information essential for interpreting the digital geo-referenced information, and the Geo-referenced Information Relations contain the actual information being exchanged.

File Identification: The **File Identification** is a fixed format file for identifying the set of data being exchanged. It is 2048 bytes long and consists of standard 7-bit ASCII characters. The fixed format facilitates the extraction of the various fields, both by computers and humans! Most of the information in the File Identification is in a free text, easy to read form (for example, the **Data Identification**, **Source** and **Maintenance Organizations**, **Copyright Statement** and **Comments**), while some is in a formatted, computer-readable form, yet still intelligible to a human (for example, the **Volume Number**, **Time** and **Date Stamps**, **Physical Record Size** and **Blocking Factor**).

The purpose of the File Identification is to allow the recipient of the data set to identify the data set, its currency and its relevance to his geographical information system, without having to do involved interpretation of the data set. The volumes of digital geo-referenced information that any user might receive, and thus the volumes of various physical media containing such information that might reside in the user's data base, are potentially enormous. The File Identification is there to provide identification of the data should the physical label on the media prove to be missing, illegible or cryptic.

In addition, the File Identification provides some information to the interface program attempting to interpret the data set - for example, the **Physical Record Size** and **Blocking Factor** indicate the manner in which the data are stored on the physical exchange medium, and the **ASCII/Binary** and **Explicit Lengths/Delimiters** flags indicate whether the data are stored using 7-bit ASCII characters or in binary, and whether the fields are separated by delimiters or whether the lengths of the fields are determined by explicit length fields appearing before each field.

The File Identification forms the **first physical file** of a data set being exchanged. The rest of the data forms the **second physical file.** On a magnetic tape, these two files are separated by two end-of-file markers. The first version of NES describes only the use of magnetic tape as the physical exchange medium, as very few users in South Africa use anything else at this stage. However, this does not preclude the use of any other exchange medium.

Global Information Section: The **Global Information Section** provides details of the data being exchanged, such as the **Projection or coordinate system** and the **Reference surface** used. Some consider this information to be information on the quality of the data being exchanged - the authors consider the information to be critical for the correct interpretation of the data being exchanged.

The entries in the Global Information Section consist of variable length fields and records with either delimiters between the fields and records, or with explicit lengths at the beginning of each field, as indicated in the File Identification. However, the use of delimiters is recommended as they are conceptually easier to understand and implement, both when creating and interpreting the data set.

Most of the entries have default values and are thus optional. Those that do not have defaults are essential, for example the **Standard meridians & parallels & scale factor.**

Other entries in the Global Information Section include the **Units** and **Increment** of the **Planimetric** and **Vertical Coordinate Resolutions**, the **Bounding Planimetric Quadrilateral Coordinate Tuples** and the **Data Quality, Feature Classification, Attribute** and **Alternate Spatial Attribute Schemes** and **Release Numbers.**

Geo-referenced Information Relations: The **Geo-referenced Information Relations** contain the actual data being exchanged. Each section, which corresponds to a table in a relational data base, contains a sequence of instances of a particular relation.

As in the Global Information Section, the sections in the Geo-referenced Information Relations consist of variable length fields and records with either delimiters between the fields and records (and sections), or with explicit lengths at the beginning of each field, as indicated in the File Identification. In addition, there is a relation, namely **TEMPLATE**, which allows the creator of the data set the option of using explicit lengths to set up templates for the fields, and hence make the fields fixed length fields. However, the use of delimiters is recommended.

The relation for exchanging information on the quality of the digital data, namely DATAQUAL, consists of free text which describes the quality of the data. In addition, the **Description** fields in the relations for exchanging the classification, namely **EXCHCLAS**, for exchanging the non-spatial attribute scheme, namely **EXCHATTR**, and for exchanging the alternate spatial attribute scheme, namely **EXCHASAS**, also contain free text. All other fields and relations contain information in a format encoded explicitly for automatic interpretation by the interface program of the recipient.

EXCHCLAS and **EXCHATTR** provide the user with a data dictionary facility for exchanging the definitions of the classification and attribute schemes together with the data set being exchanged.

Finally, there are the geometric data relations which contain the coordinate tuples and constitute the bulk of the data set - especially **CHAIDATA**, which contains the internal coordinate tuples of the chains.

Revised version of NES

The need to improve the presentation of the document, particularily the need for a user manual, has led to the revision of the published version of the National Exchange Standard. At the same time the standard itself is being revised, to:

- provide for floating point numbers,
- improve the relations for non-spatial data,
- improve the relations for raster data,
- provide for the transfer of cartographic annotation,
- include definitions of feature classes and their non-spatial attributes.

This revised version will be published as Version 2, and it is expected that it will be published in the latter part of 1990. The document will be produced so that updates can be issued as replacement pages that could be inserted in the document. This revised version is in two parts, namely Part A, giving a precise definition of the standard and Part B, which is a user manual. The user manual describes the concepts of the exchange of digital geo-referenced information, the philosophy behind NES, and how and when to use the various relations comprising NES. The user manual is intended to assist the inexperienced user in understanding the requirements of the exchange standard and its implementation.

CONCLUSIONS

The GIS user community in South Africa is currently relatively immature and small in size. This has been both advantageous and disadvantegeous to the development and implementation of the National Exchange Standard. It has been disadvantageous in that the support from the user community needed to test and comment on the standard, and to provide general expertise in the fields of application of NES has not been available. It has been advantegeous in that this standard has been established prior to the user community getting too big and thereby making the acceptance and implementation more difficult. The rapidly growing user community in South Africa will mature with NES and will undoubtably benefit from NES. In addition, the process of developing NES was the major catalyst in awakening interest in GIS in South Africa.

The intention is not to make this standard compulsory through legislation but rather to encourage its use by highlighting its advantages. The fact that the Co-ordinating Committee for the National Land Information System, which determines policy with respect to GIS for all State departments, has adopted the standard, does not mean that it is compulsory for everyone to use it.

ACKNOWLEDGEMENTS

The authors would like to thank the members of the project team that developed NES, the members of the Standards Committee and the authors' colleagues for all their help, advice and support. The opinions expressed in this chapter are those of the authors and not necessarily of their employers or the Standards Committee.

OBTAINING REPORTS

Updates and corrections of the published standard will be produced by the maintenance committee as and when appropriate. Any suggestions, comments or criticisms should be addressed to the secretariat of the committee, from whom copies of the standard are also available. The Secretariat's address is:

Secretariat: Standards Committee
c/o Surveys and Mapping
Private Bag, MOWBRAY, 7705, South Africa

The bibliography below includes all reports relating directly to NES, as well as a few describing fundamental concepts used in NES. For some of the CSIR Internal and Technical Reports listed below, copies are still available from:

Antony Cooper
CACDS, CSIR, P O Box 395, PRETORIA, 0001, South Africa
Electronic mail (UUCP): ..uunet!m2xenix!quagga!csirac
(Internet): csirac.quagga@f4.n494.z5.fidonet.org

REFERENCES

Canadian Hydrographic Service (CHS), 1985, "Evolving communications standards in the mapping and charting world: a report and a proposal", submitted to the IHO's Committee on Exchange of Digital Data.

Clarke DG, March 1989, "The National Land Information System and the Land Surveyor in South Africa", Proceedings 9th Conference of Southern African Surveyors (CONSAS 89), Vol 2, paper 9.4, 10pp.

Clarke DG, Cooper AK, Liebenberg EC & Van Rooyen MH, May 1986, "Questionnaire on GIS exchange format proposals", National standards for GIS exchange, 24 pp.

Clarke DG, Cooper AK, Liebenberg EC & Van Rooyen MH, May 1986, "Glossary of GIS terms and English/Afrikaans translations", National standards for GIS exchange, 28 pp.

Clarke DG, Cooper AK, Liebenberg EC & Van Rooyen MH, March 1987, "A proposed national standard for the exchange of digital geo-referenced information: Draft", NRIMS CSIR Technical Report TWISK 517, 151 pp.

Clarke DG, Cooper AK, Liebenberg EC & Van Rooyen MH, September 1987, "A national standard for the exchange of digital geo-referenced information", CSIR Special Report SWISK 45, 201 pp, ISBN 0 7988 3073 5.

Clarke DG, Cooper AK, Liebenberg EC & Van Rooyen MH, September 1987, "In proposing a national standard for the exchange of digital geo-referenced information", Proceedings EDIS '87 Conference, 13 pp.

Clarke DG, Cooper AK, Liebenberg EC & Van Rooyen MH, September 1987, "A Proposed national standard for the exchange of digital geo-referenced information", Final report of the project team to the National Committee for Remote Sensing, CSIR, 7 pp.

Clarke DG, Cooper AK, Liebenberg EC & Van Rooyen MH, March 1988, "On proposing a national standard for the exchange of digital geo-referenced information", South African Journal of Photogrammetry, Remote Sensing and Cartography, Vol 15, Part 1, pp 35-41.

Cooper AK, January 1986, "Report on the GIS project planning discussions", NRIMS CSIR Internal Report I668, 13pp.

Cooper AK, July 1986, "National standards for geographical information systems (GIS) data exchange", Proceedings 1st Computer Science Research Students Conference, also NRIMS CSIR Technical Report TWISK 478, 15pp.

Cooper AK, March 1987, "Thoughts on exchanging geographical information", Proceedings 1987 ASPRS-ACSM Annual Convention, Vol 5, pp 1--9, also NRIMS CSIR Technical Report TWISK 500, 13pp.

Cooper AK, September 1987, "Geographical information systems", Proceedings Computer Graphics '87, also NRIMS CSIR Technical Report TWISK 542, 10pp.

Cooper AK, December 1987, "Experiences in drafting an exchange standard", Proceedings 2nd Computer Science Research Students Conference, also CACDS CSIR Internal Report I800, 12 pp.

Cooper AK, September 1988, "A data structure for exchanging geographical information", Quæstiones Informaticæ, Vol 6, No 2, pp 77-82, also NRIMS CSIR Technical Report TWISK 535, 14pp.

Cooper AK, September 1988, "Exchanging geographically referenced information - a status report", Proceedings Computer Graphics '88, pp B1-6 - B1-20, also CACDS CSIR Technical Report PKOMP 88/17, 12pp.

Cooper AK, April 1989, "The South African standard for the exchange of digital geo-referenced information", Proceedings Auto Carto 9, pp 745-753, also CACDS CSIR Technical Report PKOMP 89/10, 11pp.

Cooper AK, 1989, "A survey of standards for the exchange of digital geo-referenced information", South African Journal of Photogrammetry, Remote Sensing and Cartography, Vol 15, Part 3, pp 136-140, also CACDS CSIR Technical Report PKOMP 89/7, 8pp.

Cooper AK & Scheepers CF, April 1989, "Classification of digital geographical information", Poster paper at Survey and Mapping 89, University of Warwick, also CACDS CSIR Technical Report PKOMP 89/9, 13pp.

Digital Cartographic Data Standards Task Force, January 1988, "The proposed standard for digital cartographic data", American Cartographer, Vol 16, No 1.

Greenwood PH, September 1988, "Using the proposed national exchange standard for GIS data", Proceedings Computer Graphics '88, pp B1-21 - B1-29.

Haywood P, chair, 1986, "The National Transfer Format final draft (issue 1.3)", Ordnance Survey, Southampton, 158pp.

Lane A, 1988, "Book review: A national standard for the exchange of digital referenced information", International Journal of Geographical Information Systems, Vol 2, No 1, pp 81-82.

Moellering H, ed, 1985, "Digital Cartographic Data Standards: an interim proposed standard", Report no. 6, National Committee for Digital Cartographic Data Exchange Standards, 219pp.

National Exchange Standard Committee, July 1988, "Questionnaire: National Exchange Standard", Chief Directorate: Surveys and Mapping, 11 pp.

North American Work Group Committee on the Exchange of Digital Data (NAWG), 1985, "Proposed format for the exchange of digital hydrographic data", submitted to the IHO's Committee on Exchange of Digital Data.

Scheepers CF, Van Biljon WR & Cooper AK, September 1986, "Guidelines to set up a classification for geographical information, NRIMS CSIR Internal Report I723, 12 pp.

Standards Association of Australia (SAA), 1981, "Interchange of feature coded digital mapping data", Australian Standard 2482-1981, 24pp.

Van Biljon WR, March 1987, "A geographic database system", Proceedings Auto Carto 8, pp 689-700.

Van Biljon WR, September 1987, "Towards a fuzzy mathematical model of data quality in a GIS", Proceedings EDIS '87 Conference, 11pp.

Van Roessel JW, 1987, "Design of a spatial data structure using the relational normal forms", International Journal of Geographical Information Systems, Vol 1, pp 33-50.

Van Rooyen MH & Boerstra J, September 1987, "Topology in geographical information systems", Proceedings EDIS '87 Conference, 14pp.8

THE SWEDISH STANDARDIZATION PROJECT WITHIN THE FIELD OF GIS

Clas-Göran Persson
National Land Survey
S-801 82 Gävle, Sweden

Torbjörn Cederholm
SIS-STG
Box 3295
S-103 66 Stockholm, Sweden

ABSTRACT

This chapter describes the Swedish Standardization Project concerning the exchange of geographic data and other closely related areas within the field of GIS. The planning and setting up of the project is also described and experience gained from work carried out to date is presented. The project s main objective is to adopt an existing internationally accepted transfer standard. There are, however, still many questions which need to be solved at national level. A detailed description of the conceptual platform is dealt with as well as influences by other countries.

INTRODUCTION

Cartography in Sweden, as in many other countries around the world, has changed considerably during the last few decades. Above all, the development within the field of computer technology has meant that many parts of cartographic work have been automated and made more efficient.

The first steps towards current digital cartography were taken as early as the mid-sixties through the introduction of numerically operated drawing equipment. During the seventies, this technology was gradually improved and, as time elapsed, became increasingly advanced. However, it was still a question of using computer technology for the production of traditional maps, i. e. the ultimate aim was a draughted or printed product.

A new phase of development took place towards the end of the seventies and beginning of the eighties with the building up of cartographic data bases. These data bases could then be used for several different products, which meant improved efficiency both in connection with the production of new maps as well as keeping maps up-to-date. Even so, the work was dominated by a strict cartographic way of thinking. Presenting information on maps and map production was still the main objective.

Today, Geographic Information Systems (GIS) and Geographic Data Bases are the terms heard most, which has meant that pure cartographic applications have to some extent been given less of a leading role. The map is no longer the ultimate aim but used rather as an aid to other, in many cases more important, applications.

There is, naturally, a certain need for data exchange between **cartographic data bases.** The motivation behind the building-up of these data bases, however, has been provided by organizations endeavouring to rationalize their own particular work.

With regard to data exchange between **geographic data bases**, on the other hand, the needs are more accentuated. The whole concept is based on the idea that each organization is responsible for its own data and that other interested parties can gain access to this data as and when the need arises.

SCOPE AND GENERAL GOALS

Overall Aim

It is above all the development of GIS, therefore, that has lead to a demand for standards for data exchange. The Swedish Standardization Project aims to fulfil these needs, i. e. to facilitate the exchange of geographic data in digital form.

Cartographic Data vs. Geographic Data

How exactly can "cartographic data" be distinguished from "geographic data"?

A somewhat simplified explanation is that **cartographic data** describes a model of a map, which in itself is a model of the real world, whereas geographic data constitutes a model of certain aspects of the real world without reference to a map.

In addition cartographic data aims at presentation whereas **geographic data** aims at analysis (see figure 1), which sometimes means a slight conflict.

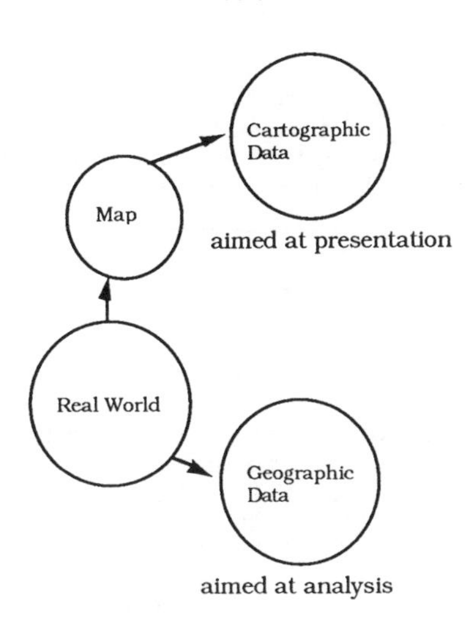

Figure 1. Cartographic vs. Geographic Data

Also, cartographic data is often in the form of spaghetti-data while geographic data usually has a more complex structure which includes topology.

Scale in Digital Cartography

Scale is a concept which could benefit from analysis in connection with digital cartography. One of the most important arguments put forward in support of the change to digital technology was surely the desire to eliminate scale dependency. The fact is, however, that the question of scale is still relevant.

Small-scale maps have previously been used for applications requiring an overview. Their digital counterparts could be called small-scale cartographic data bases in which a wealth of detail gives way to a broader and more comprehensive view.

In large-scale maps and large-scale cartographic data bases in principle the opposite is true. A broad, comprehensive view gives way to a wealth of detail.

Therefore, when we talk about scale in connection with cartographic data bases we do not mean geometric scale. In this respect, the data bases are, of course, independent of scale, as co-ordinates are usually stored in the geodetic system. Instead, we associate scale with the selection and degree of detail and therefore the type of applications for which the data is meant to be used.

Our long term ambition is to fulfil the need for nationwide standards covering all scales.

Areas of Interest
The main aim of the work done within standardization is to facilitate the **exchange** of digital geographic data. In order for this to be possible, certain principles must be uniform, which in turn will influence conditions in the data bases themselves.

The work covers the following main areas:

- o Terminology and Basic Concepts in GIS
- o Definition and Classification of Geographic Entities and Attributes
- o Data Structures and Transfer Formats
- o Geodetic Reference Systems
- o Quality Description of Digital Geographic Data

Work is in progress within all of these areas. Work within Geodetic Reference Systems, however, is carried out and managed solely by the National Land Survey and is therefore not part of the project as such.

As can be seen, providing standards for data exchange is just one of many problems to be solved.

What can be learnt from other countries?
Sweden is not the only country working on the problems of standardization within this field, many other countries are, in fact, seveal years ahead in their work.

So in order not to reinvent the wheel it is important to take advantage of the work already done in these countries. For this reason, the aim is to adopt an internationally accepted transfer format rather than to develop our own. There are, however, still many questions needing to be solved on a national basis.

HISTORY AND BACKGROUND

Earlier Efforts
In Sweden, no concentrated standardization effort within this area has really taken place before. At the beginning of the eighties, however, the Swedish Association of Local Authorities (Kommunförbundet) initiated the development of a transfer format for cartographic data which also included a catalogue of objects used in large-scale applications.

This format - known as **KF85** - has gradually acquired the character of de facto standard for data exchange in Sweden. The data structure is, unfortunately, too elementary to be able to meet the demands made by current GIS applications. Moreover, it can only handle vector data, not raster and attribute data.

Project History
The increasing need for data exchange and co-ordination of

the handling of geographic data eventually led to the only possible solution - a national standardization effort.

It can be said that the project came about with the establishment of **The R & D Council for Land Information Technology (ULI,** the Swedish acronym for Utvecklingsrådet för Landskapsinformation), in October 1986. ULI is a membership organization open to users, producers, and researchers from government and local authorities, R & D institutions, and private companies. Its main role is to assist in the coordination of R & D activities concerning the use of geographic data and the development of Geographic Information Systems.

Standardization work fits in well within this role and the new organization provided the natural platform for this type of work. ULI immediately began its work on standardization and during 1987 a study was made of similar work in other countries, especially work being carried out in the UK and the US.

The next break-through was professor Moellering s visit to Sweden in February 1988. During a two-day seminar he gave an extensive presentation of the American standardization efforts. This saved a lot of time: the work could get started more quickly and be organized in a more suitable way. Despite this, it took more than 10 months of preparatory work to get the project launched, and the official start finally took place in January 1989.

The work done during the first few years has been something of a trial period: Questions concerning finance had not yet been solved, the structure of the organization was the subject of perpetual discussions, people from different organizations with different backgrounds had to be encouraged to work together etc., etc.

There was also a great need for training and education. This concerned partly the standardization work itself but there was also a need for a broader knowledge of modern GIS technology. A number of courses and seminars were arranged in order to satisfy these needs, but a great deal of time was also spent in the working groups discussing questions of a more basic nature. Looking back though, we can say that this phase was in fact necessary in order for work to progress.

A get-together-seminar held in Norrköping in February 1990 meant something of a new start and enabled the project to get off the ground. The overall goals were finally decided upon and detailed plans for the project were drawn up and checked against one another. In this way, all participants were given a common ground from which they could continue the work.

DESCRIPTION OF THE CURRENT EFFORT

Conceptual Platform
It has been decided that a Swedish GIS terminology should be provided and also that a conceptual model for the handling of geographic data in Geo-graphic Information Systems should be produced.

A conceptual model of this kind is usually referred to as a **Meta Model**. In such a model, basic concepts and their relationships are defined and given appropriate terms. The most important parts of the Meta Model are dealt with in the following description.

There are two stages involved in producing a model of the real world in a GIS:

- First a model is created which constitutes a well-structured and to a certain extent simplified picture of phenomena in the unstructured real world. The building blocks in this **Real World Model** are called **entities**.

- With this as the starting point a **Data Model** is then built. This describes how the Real World Model is to be represented in digital form. The Data Model consists of **objects** and forms the basis for the construction of a data base.

As a result, the structure as shown in figure 2 is obtained. Here the relation between the entity and the object is clear: an object is a digital representation of all or part of an entity.

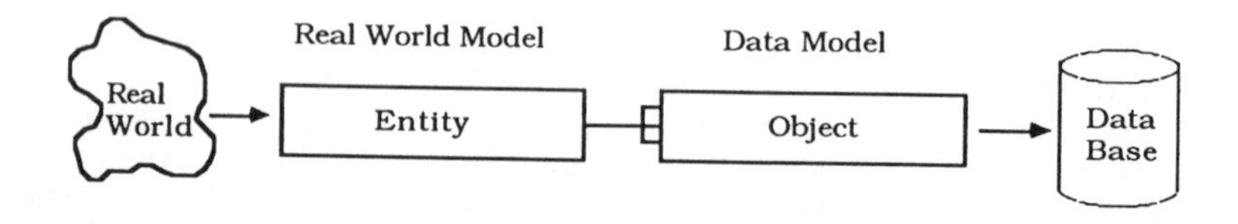

Figure 2. Different phases in producing a model of the real world.

The characteristics of real world phenomena are described by **attributes** and **relationships** connected to entities or objects - depending on in which model you find yourself . This approach is for the most part in accordance with the American standardization work although certain diffe-rences are apparent, after a more detailed analysis is carried out.

In the Swedish version, objects are built up by **geometric elements** and **attribute elements** as shown in figure 3.

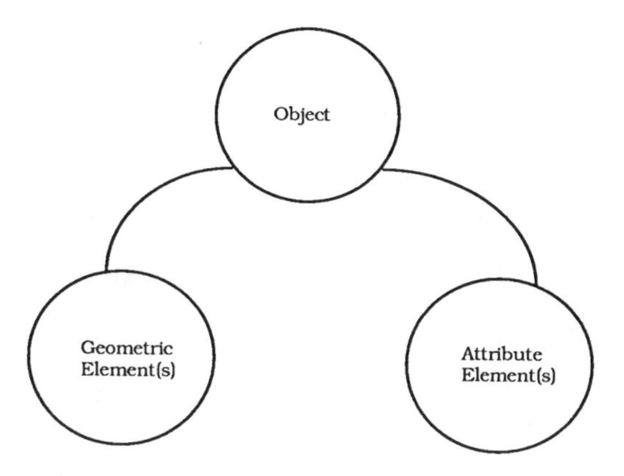

Figure 3. Construction of an object.

Consequently, there is a difference between the object as such and its geometric description. Such a distinction is not made in the American standard, but this way of tackling the problem has certain advantages and is more general by nature. However, the result after implementation of the data model is often the same, at least in current GIS.

The reasons why such a high priority is given to terminology work and the development of a conceptual model are the following:

- Firstly - and perhaps most important - there is a need to create a greater understanding of the relationships between, and the meaning of, the different concepts that are introduced. A conceptual model can help to provide this.

- Secondly, a conceptual model can provide the universal and neutral framework needed to ensure compatibility between different parts of the standard that are being developed within different working groups.

- Thirdly, developing a conceptual model in itself gives the participants a better insight into the complexity and nature of the task at hand.

- Finally, a common and consistent terminology not only has an important role within the project itself, but is also vital when it comes to introducing this new technology.

Ultimate and Immediate Goals
Both ultimate and immediate goals have been defined within the project.

As has been stated earlier, the ultimate goal is to adopt an internationally accepted transfer standard for digital geographic data. This standard should cover **geometric** as well as **attribute** data and include both **vector** and **raster** representation.

The ultimate goal includes therefore the task of analysing the various candidates for this standard, as well as dealing with the other main areas as described above.

The more immediate goals are:

- o to develop a simple raster data transfer standard for mass transportation of remote sensing data etc., and

- o to implement a preliminary data quality description standard into the de facto transfer standard (KF85) mentioned above.

Organization
The project is organized in the following way:

- o **The Swedish Standards Institution** (SIS-STG), is responsible for the actual standardization procedures. This makes sure that formal standards are obtained.

- o A **Steering Committee**, with representatives from the most important organizations and financial contributors, has the overall responsibility.

- o A **Technical Committee** manages and coordinates the every day work, carried out in **Working Groups** - one for each part of the standard.

In addition to that, several bodies - **Reference Groups** - are associated to the project, to which committee reports and draft standards are sent for consideration and comments.

ULI participates at several of these levels.

Working Groups
Within the GIS-related area four working groups are now active. These are described below. In addition, a fifth working group (**Wg II**) has been for-med to deal exclusively with **GPS Terminology**. More working groups will be established later on. Terminology work is placed directly under the Technical Committee.

<u>Entity Classification (Wg I):</u> This working group is responsible for structuring and classifying geographic entities and their attributes. The approach is in principle

the same as that of similar projects in the US, the UK and West Germany, and work has begun with entities in the utility sector (water, electricity, telephone etc.).

Digital Field Survey Systems (Wg III): This working group will develop a transfer standard for digital field survey data. This involves both the communication between total stations and field computers and between field computers and other processing systems. The aim is to achieve manufacturer independent data transfer in both directions, with respect to communication protocol, transfer format, and information content. The resulting transfer mechanisms should be compatible with the overall Data Model.

Raster Data Transfer Format (Wg IV): To begin with, this wor-king group will develop the standard for magnetic tape transfer of raster data. After that, the group will join the more long term effort to adopt a complete standard for digital geographic data - including vector, raster as well as attribute data.

Data Quality (Wg V): The task of this working group is firstly to revise the existing draft version of the standard for data quality description, thus making it workable , secondly to implement this preliminary version into KF85, and finally to further develop the quality description concept to a formal standard.

The Data Quality Draft Standard follows in the main the approach used in corresponding American and British standards. The concept allows for varying levels of complexity as well as flexibility concerning the diffe-rences in the technical implementation.

- The basis of all data quality description is the creation of a **Data Base Specification**, which is a general description of the main characteristics of a data set. This also contains a **Quality Report**.

- The actual data quality description is done under the headings: **lineage**, **accuracy**, **currency**, and **completeness** (including **logical consistency**).

- Description can be done on **different levels** - from an object level, which is the lowest, to common information for the whole data set. In addition, a differentiation is made between **physical description** i.e. explicit within the data, and **separate description** i.e. in the Quality Report.

Influence from Other Countries
Our main influences are: The ATKIS project (West Germany), SDTS (US), NTF (UK), and the work carried out within CERCO Working Group V.

- The West German **ATKIS-project** has above all influenced the choice of structure for the Data Model and the method for structuring geographic data used by Wg I.

- The American standard SDTS (**Spatial Data Transfer Format**) has mainly influenced terminology and the creation of the Meta Model.

- The British transfer format NTF (**National Transfer Format**) is one of the candidates for the Swedish national transfer standard.

- CERCO is an interest organization for Western European mapping authorities. The Swedish representative is the National Land Survery. **CERCO Working Group V** (Wg V), which deals with standardization, has recently put forward a very interesting idea which is to package the NTF in the ISO8211-format in such a way that it would be compatible with the ATKIS data model. This ISO8211/ATKIS-compatible variation of NTF provides the main candidate for what within CERCO is called ETF (**European Transfer Format**). Such a solution would also be a very interesting alternative as far as Sweden is concerned. (The work within CERCO Wg V is described elsewhere in this monograph.)

Project Status

As previously mentioned, the project, after a period of trial and error , has now entered a more expansive stage. All the working groups have begun their work, but no complete standards have been produced yet.

The future looks bright, however. Amongst other things, the Swedish Government has provided the means for work on standardization to continue, which has enabled a special secretariat be set up to deal with the day to day running of the project.

Financing, in the introductory phase of the project, has been provided mainly by participants contributing with their working hours. The new arrangement, from the start of the financial year 1990/91 (July 1990), means instead that the Government now provides financial support in proportion to the amount of money contributed by interested parties.

The following time schedule has been decided for the GIS related working groups:

- **Working Group I;** a preliminary version of an entity catalogue for the utility sector - autumn 1990.

- **Working Group III;** a draft standard - autumn 1991, a final standard by the turn of the year 1993/93.

- **Working Group IV;** a draft standard - autumn 1990.

- o **Working Group V**; a quality description included in KF85 - spring 1991.

Experience
The experience obtained so far shows that the limiting factor in standardization work is not only money, but also people. Sufficient funding of the project is of course important, but the main problem is to find people that are willing/able to spend time on the project. Most working group members have to combine standardization work with their ordinary work, with no reduction of the latter.

It is also clear that standardization work within the field of GIS is very much linked with R & D activities. This means that the people needed have to be experts on some aspect of GIS, which in turn reduces the number of available persons even more. It also implies that one has to take part in the necessary R & D activities - or wait for others to produce the answers.

This frontline nature of the work means of course that many of the users are some years behind in terms of understanding the different concepts and ideas involved. It is important, therefore, to spend a lot of time on information and, in the future, training activities.

CONCLUSION

After a somewhat laborious "apprenticeship", work on the project has begun in earnest. All five working groups have made good progress and although no final standards have been produced as yet, results are expected from all groups within the next two years.

Therefore, the future looks bright. The Government has recognised the importance of this work. Also, the conditions necessary to bring about an international - or at least European - transfer standard appear to be good.

The tentative steps taken in the initial stages of the project must, in retrospect, be seen as a necessary part of the development process. So, the conclusion, in short, is that standardization work within the field of GIS is interesting, demanding but takes a long time. It should be appreciated that standards do not live for ever. Time and resources must be made available in order to maintain and revise standards in the future.

CONTACT

At the time of writing (August 1990), no actual technical reports or final standards have been produced within the project. Existing material is for the main part in the form of preparatory papers. Articles which have appeared in periodicals and journals, concerning the project, are listed in the references below.

In accordance with the time schedule for the project, a number of reports and standards will be presented in the near future. This material can be obtained through SIS-STG or ULI from the following addresses:

SIS-STG
Box 3295
S-103 66 Stockholm, Sweden
phone: +46 8 613 52 00
fax: +46 8 14 93 92

ULI Secretariate
S-801 82 Gävle, Sweden
phone: +46 26 15 30 80
fax: +46 26 68 75 94

Questions of a technical nature can be directed to Torbjörn Cederholm at SIS-STG (see above), co-author of this chapter, and project leader as from 1 September 1990.

REFERENCES

Arbeitsgemeinschaft der Vermessungsverwaltungen der Länder der Bundesrepublik Deutschland (AdV) - Arbeitsgruppe ATKIS, 1989, ATKIS-documentation, Status 01.09.1989, Bonn, FRG: Landesvermessungsamt, Nordrhein-Westfalen.

Cederholm, T., and Persson, C.-G., 1989, "Standardiseringsverksamheten i Sverige inom GIS-området - terminologifrågor, informationsstrukturering och kvalitetsmärkning", Geografiska informationssystem - föredrag vid ULIs utbildnings- och informationsdagar 1989, ULI- rapport 1989:4, pp. 27-39.

Olsson, O., and Persson, C.-G., 1988, "Förslag till standard för kvali-tetsmärkning av digital lägesbunden information", ULI-Information, 1988:1, pp. 21-27.

Spatial Transfer Standard Technical Review Board, 1990, Spatial Data Transfer Standard, Reston, VA: U.S. Geological Survey, 181 pp.

Working Party to Produce National Standards for the Use of Digital Map Data, 1989, National Transfer Format, Release 1.1, Southampton, UK: The NTF Secretariat, Ordnance Survey.

GIS/LIS DATA EXCHANGE STANDARDS: ACTUAL SITUATION AND DEVELOPMENTS IN SWITZERLAND

Christoph Eidenbenz, dipl.Ing.ETH
Federal Office of Topography
Seftigenstrasse 264
CH-3084 WABERN, Switzerland

INTRODUCTION

Since the early 1980s when Geographical Information Systems or Land Information Systems (GIS/LIS) became available for practical applications, the responsible organizations in Switzerland got more and more involved and the number of installed GIS/LIS is increasing from year to year.

During the acquisition phase, the buyer of a GIS/LIS generally concentrates on his specific needs and possibilities and has often no interest in coordinating his activities with other organizations.

In the acquisition of GIS/LIS, the organizations responsible for the management of geographical data in general proceed according to the classic project steps:

1. Evaluating and selecting a hardware/software system

2. Installation and training

3. Digitizing the existing graphic documents

A takeover of existing data is not possible since the basic information is available only in an analog form. In most cases, unfortunately, the need for an appropriate general input/output format does not become apparent until phase 3. At than point the user largly depends on the system supplier. Standards can be helpful if they are supported by the main system suppliers but very often the input/output format of one or two main systems becomes a defacto standard.

The political and administrative structures in Switzerland are fairly complex. To understand the actual GIS/LIS situation in our country, the impact of this structure on the different GIS/LIS applications must first be explained.

The decision-making authorities and the corresponding administrations in most of the sectors are a hierarchy organized on three distinct levels:

- o Federal level
- o Cantonal level
- o Communal level

Depending on the task, the decisions, financing and execution are attributed to one or more levels. Certain tasks, for example cadastral surveying, are executed in the private sector and controlled and financed by authorities of all 3 levels. Other tasks such as telecommunication is centralized within the Swiss PTT.

The federal authorities are responsible for the tasks with respect to the country as a whole. The activities are concentrated in the legislative, financial and supervisory branches, leaving the actual execution of the tasks mostly either up to the 26 cantons, the 3022 communities or to the private sector.

The cantonal authorities have a rather strong position since they are financially independent (having the right to levy taxes). For certain tasks, they are bound by federal legislation. On the other hand, they are subsidized by federal authorities and have certain liberties in executing their tasks.

At the communal level, the executives accord themselves to the federal and cantonal laws and prescriptions. There are nevertheless vast differences between the well-organized large communities of cities like Zurich, Bern and Geneva and the very small mountain communities with some 20 to 50 taxpayers.

Table 1 shows the distribution of typical GIS/LIS tasks and their relation to the different levels.

ACTUAL SITUATION AND PROJECTS

Topographic maps and geodetic control

The Federal Office of Topography is an independent division of the Department of Defense and as such responsible for the geodetic control (1st to 3rd order triangulation) and the topographic mapping of Switzerland. It serves as a military as well as a civil mapping agency of the country.

TABLE 1
GIS/LIS projects in Switzerland

	Federal level	Cantonal level	Communal level	Systems used
Topographic mapping	Federal Office of Topography Project DIKART	--	--	SCITEX, graphics art-oriented systems
Cadastral survey	Federal Direcorate of Cadastral Survey, supervision Project RAV	20 Cantonal Cadastral Survey supervision	350 licensed surveyors execution	AM/FM systems Intergraph SICAD, GRADIS, INFOCAM etc.
Planning	Federal Planning Office supervision	Cantonal Planning Offices	Communal Planning Boards supervision Private planners execution	ARC-INFO
Statistics	Federal Office of Statistics Project GEOSTAT : Landuse Statistics			ARC-INFO
Utilities:				
Telecommunication	PTT with 18 regional offices Project TERCO-GRAFICO	--	--	SICAD
Electricity	supervision	Priv. companies with financial participation by the cantons	Private companies with financial participation by the communities	diverse AM/FM systems
Water	supervision	supervision	Communal water administration	diverse AM/FM systems
Railways	Swiss Federal Railways with 3 regional offices Project : **DfA**	--	--	Intergraph

The information content of topographic maps consists essentially of the geometry of the chosen map elements and their rather few attributes. Therefore, a topographic information system can be realized in a fairly straightforward manner. On the other hand, the basic framework of the geographic data for many users is given by the geometric data of the topographic map. But since the geometric structures are somewhat simple and the user's input/output formats are well-known, our Office has adopted an open policy in controlling and using the main transfer formats instead of developing its own standard.

The Swiss National Map Series were completed in 1979. They include 249 sheets at the scale 1:25,000, 78 sheets at 1:50,000, 23 sheets at 1:100,000, 4 sheets at 1:200,000 and one sheet each at 1:500,000 and 1:1,000,000. All of these typically topographic maps have a very high accuracy standard and a high information content.

Since 1968, the maps are being systematically revised in a 6-year cycle using aerial photography as source material and stereocompilation to extract the new information. After the National Map Series had been completed, their revision has become the main task of our Office.

In 1982 a first project (Project DIKART) in the area of digital cartography was started. The main goal was to establish a digital height model (DHM) of Switzerland using the map information, i.e., contours and spot heights, as a primary input. The necessary equipment (SCITEX) was acquired with respect to future applications in digital mapping and map revision.

As an additional product, a multi-color rasterized digital map data set (pixel map) at the scale 1:25,000 with a resolution of 50 dots per inch resulted.

For the last 2 to 3 years, the demand for vector data has been increasing. As a consequence, work is under way to digitize the small scale maps 1:1,000,000, 1:500,000 and 1:200,000. The trend is obviously also towards vector data in larger scales.

The information content of topographic maps consists essentially of the geometry of the chosen map elements and their rather few attributes. Therefore, a topographic information system can be realized in a fairly straight-forward manner. On the other hand, the basic framework of the geographic data for many users is given by the geometric data of the topographic map. But since the geometric structures are somewhat simple and the user's input/output formats are well-known, our Office has adopted an open policy in controlling and using the main transfer formats instead of developing its own standard.

For the next decade the principal goals in digital cartography at our Office are:

1. Completing and refining the DHM of Switzerland

2. Creating a topographic database with an accuracy of 1 meter

3. Solving the problem of digital map revision in all scales with a hybrid raster/vector technique

Cadastral survey
The cadastral survey in Switzerland is in general executed by some 350 private licensed surveying offices. Each office operates in a specific region and retains a kind of monopoly there. This structure and the monopoly situation differ from canton to canton. At present, the surveys at the scales 1: 1,000 (towns 1:5,00) are predominantly in a graphic form. In newer surveys, the lot boudaries are also registered numerically in computer files and the most recent surveys tend to be fully numerical. The overwhelming part of the cadastral survey is nevertheless still graphic.

The work of the licensed surveyor is controlled by the cadastral administration office of the canton. These offices have an average of 10 to 20 employees. Their main tasks are:

- o controlling and testing the on-going surveys and the updates, verifying the content and the pre-cision according to the federal prescriptions and tolerances

- o coordinating and providing the finances (federal, cantonal and communal subventions)

- o maintaining the 4th order triangulation under the supervision of the Federal Office of Topography

- o maintaining the general cadastral plan (1:10,000/5,000)

The Federal Directorate of Cadastral Survey (Eidg. Vermessungsdirektion) has the overall management and control function. It coordinates the different activities in the cantons and the communities and is responsible for the allocation of federal subventions. It defines the technical standards and the precision tolerances.

In the cadastral field there is a tremendous movement towards land information systems. In an overall reform project called RAV (Reform Amtliche Vermessung) headed by the federal office, the graphic cadastral plans are to be transformed into a numerical form and replaced by an appropriate LIS structure.

The RAV project
The RAV project was started in 1978 with the main purpose of transforming the existing graphic cadastral plans and expanding them into a land information system. Since the organization of the cadastral survey in Switzerland is very complicated and decentralized, the project is advancing rather slowly. At present, it is in the decision-making phase (financing and execution). Looking at the time schedule, it should take another 20 to 30 years to complete the project.

Technically, the RAV project is based on 3 essential elements:

- o layer model with 11 layers
- o data catalog
- o data exchange standard

Of the 11 layers in the model, 3 are optional. If an optional layer is chosen in the execution, the technical form of the model must be accepted. The present layers are as follows:

1	Control points	mandatory
2	Soil coverage	mandatory
3	Single objects, line elements	mandatory
4	Names	mandatory
5	Property boundaries	mandatory
6	Property registration	optional
7	Planning zones	mandatory
8	Utility lines and mains	optional
9	Heights	mandatory
10	Land use	optional
11	Administrative division	mandatory

The horizontal accuracy and the number of details to be captured in the different levels depends on the location of the areas and is divided into 5 classes:

I + II Built up areas (cities, center of towns)

III Agricultural areas

IV + V Mountainous areas

In the data catalog, the items of the different levels are described in detail togehter with their format. Some examples are given in Table 2.

The data catalog is not yet finalized. It was established with organizations and administrations in the neighbouring fields of interest but at the moment there is very little experience as to whether the choice covers the (as yet unknown) demands of the potential users.

The data exchange interface AVS (Amtliche Vermessungs Schnittstelle) is foreseen as the official data exchange standard between the different organizations in the cadastral survey. It should also be used for delivering data to external users and for the regular backup of the LIS data.

In July 1990, a first version was presented by the RAV project management. The standard is based on the described data catalog and consists of 2 elements:

- o the data description language INTERLIS

o the transfer format description

TABLE 2
Examples of object classes with data format

Object-class	Requirements	Field
BUILDING :		
Identification	insurance no.	TEXT*12
Name	important name	TEXT*30
Coordinates X,Y	in cm	999999.999
Quality X,Y	quality measure	9/9.9
Number of dossier	given by Canton	TEXT*12
Validity	existing/project	X
Date	last treatment	JJJJ.MM.TT
Geometry	vector/left/right	
LINEAR ELEMENT :		
Identification	to be defined	TEXT*12
Coordinates X,Y	in cm	999999.999
Quality X,Y	quality measure	9/9.9
Number of dossier	given by Canton	TEXT*12
Validity	existing/project	X
Date	last treatment	JJJJ.MM.TT
Geometry	vector	
TYPE :	narrow river	X
	narrow path	X
	powerline	X
	rail	X
	skilift	X
	etc.	
Total: 32 categories in 11 levels		

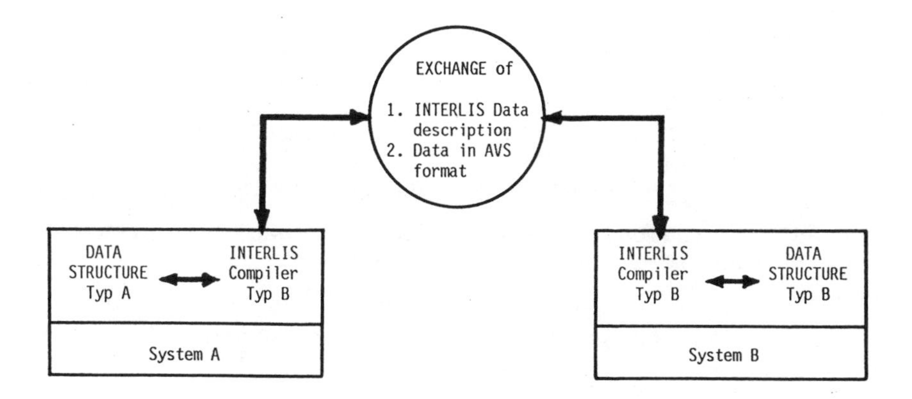

Figure 1. Data Transfer

The data exchange mechanism is shown in Figure 1. INTERLIS describes the data in an appropriate logical scheme (extended Backus-Naur notation). On every LIS accepting this standard, the system supplier has to write a software interface for the transformation of the data change format into the system's internal data structure according to the data description in INTERLIS notation.

It is too early to make a statement on the acceptance of this standard. Since it is supported by the RAV, the firms will probably supply the necessary interface software in the near future. The standard is nevertheless rather complicated for the exchange of simple items.

Planning

In the planning sector, the development is rather slow. There are several ARC/INFO systems in the private sector, used mainly for environmental planning and inventories. The urban planners tend towards RAV-compatible systems since the urban development plans are based on property boundaries. Some cantonal and federal administration have database systems with graphic image display capabilities used for the regular registration and supervision tasks.

Statistics

Under the project called GEOSTAT, the Federal Office of Statistics is working on a land use information system based on a grid structure of 100 x 100 meters (hectar-grid). The data is captured by stereo-interpretation of aerial photographs taken in the map revision program by the Federal Office of Topography. A specially distorted hectar-grid overlay allows the appropriate localization of the grid cell. It is planned to also incorporate linear elements into GEOSTAT. The data will most probably be delivered by the DIKART project.

Telecommunications

In telecommunications, the Swiss PTT (post, telephone, telecommunication) retain a monopoly. The PTT network is graphically mapped in two different forms:

- Network plans indicating the exact geometry of the network on the basis of cadastral plans and including information regarding line and node qualities.

- Topological schematic plans indicating detailed technical information on the nodes and links of the network.

Actually, these plans are established and maintained in 18 regional offices, the cadastral information coming from the licensed surveors.

The PTT have a centralized structure and are very well organized. They started in 1985 with a project to transfer their network information into an LIS. The project, called TERCO-GRAFICO, is well under way. Siemens SICAD hardware and software is used to manage the task. Since the cadastral information is not available in digital form, the PTT have started digitizing the cadastral plans together with the geometry of the network as a first step. In a future phase, cadastral information will be generated by the cadastral organization and taken over by the PTT. Since the planned cadastral data transfer standard is not yet available, the PTT ordered the system supplier to deliver format transfer software to read the formats of the main data suppliers.

In the future, the PTT plan to take over the revision data of the cadastral survey. They are therefore interested in the AVS standard for communicating with the world of cadastral surveying.

Electricity
Electricity management is controlled by the private sector. There are several large private companies producing hydroelectric and nuclear power and handling their distribution in a vast powerline network. The local distribution is also managed by the same companies except in larger cities and villages. The cantons and cities participate as shareholders and have a certain influence on management decisions.

The situation here is essentially the same as with the PTT. The companies tend to concentrate on the digital capture of the network-data and wait for the cadastral data. If the reform is not adopted and the data not delivered within a reasonable time limit, the companies will take the already digitized property boundaries and other planimetric data over from the PTT, or start digitizing on their own. The electricity companies, like the PTT, are very much interested in the AVS standard in order to take over either new or revised cadastral data.

Water management
The water management and distribution is organized on the community level. Most of the cities and villages have their own water administration. Except for the fact that the management here is taken care of by the administration, the situation is the same as with the electicity.

Federal railways
The railway administration is completely independent as far as the survey of their property is concerned, i.e., property boundaries and rail installations. They have recently started an LIS project called DfA (Datenbank der festen Anlagen = database of the fixed installations). To construct and maintain their installations, the federal railways traditionally contract their projects to private engineering companies. Therefore, they have a keen interest in a data exchange standard.

CONCLUSIONS

In Switzerland, there is actually a two-fold situation:

On one hand, we have the cadastral community with a large number of applications and users. Here we find a reasonable data exchange format within the RAV project. The crucial question is: Will the RAV project management have enough influence to impose the format on the system suppliers and the private surveyors? If so, this standard will enjoy wide acceptance.

On the other hand, there are users working with less accurate data but covering uniformly larger areas. In this field the users have to live with the input/output format of the systems they work with. The standard in this area could well be a European standard either developed by CERCO or taken over from a leading organization. Good examples would be the National Transfer Format of Great Britain or the NATO standard called DIGEST.

Considering the complicated structure in the GIS/LIS environment, we will always have to live with more than one data exchange format, even in a small country like Switzerland.

REFERENCES

Eidg. Vermessungsdirektion, 1987, "Reform Amtliche Vermessung RAV, Detailkonzept" (German, French and Italian version). Distribution: Eidg. Drucksachenund Materialzentrale, Bern.

Eidg. Vermessungsdirektion, 1990,
a) "Datenkatalog mit Anforderungen"
b) "INTERLIS, ein Datenaustausch Mechanismus fur Landinformationssysteme"
c) "AVS Beschreibung des Grunddatensatzes in INTERLIS"
Distribution: Eidg. Vermessungsdirektion, Eigerstrasse 65, Bern.

Eidenbenz, Ch., 1987, "Aufbau einer digitalen Karte", Series Krieg im Aher, Vol.XXV,. Bundesamt fur bermittlungstruppen, Bern.

Gnehm, B., 1990, "AM/FM fur die Bewirtschaftung des PTT Leitungsnetzes", Proceedings AM/FM Conference, Montreux, Switzerland.

Bundesamt fur Statistik, 1990, GEOSTAT, eine raumbezogene Datenbank des Bundesamtes fur Statistik", Bulletin: Info a la carte. Distribution: Bundesamt fur Statistik, Servicestelle GEOSTAT, 3000 BERN.

Zeitschrift, 1990, Vermessung Photogrammetrie Kulturtechnik, "Spezialnummer RAV".

THE NATIONAL TRANSFER FORMAT

M. Sowton
Chairman NTF Steering Committee
Ordnance Survey
Romsey Road, Maybush, Southampton, SO9 4DH

ABSTRACT

This Chapter describes the way in which the National Transfer Format was created and its subsequent development as an exchange format for digital map data. The concept of using different levels to allow flexibility when exchanging complex data sets will be maintained in new proposals to make NTF compatible with ISO 8211 (ISO, 1985).

INTRODUCTION

In Great Britain the issue of a transfer standard for digital map data was addressed by the Ordnance Survey (OS) as the main investigator and supplier of such data. It was recognised as early as 1969 that in order to transfer map data to customers it would be essential to create a format that would allow the internal formats used in the OS digital system to be converted to a simple transfer format universally understood and accepted by customers for OS digital data. This early format known as DMC was used until 1984 as the format in which OS transferred digital map data to its customers.

In 1984 DMC was modified to allow the transfer of more sophisticated data and a new transfer format called OSTF was introduced for all data of this type. Independently the National Joint Utilities Group (NJUG) adopted this transfer format with some enhancements for the transfer of data between utilities and OS made some changes to the header to allow greater flexibility.

However not all customers were able to use the new format specification and both DMC and OSTF are still available, although DMC has almost entirely been superseded by OSTF, and OSTF will be phased out during 1992.

It was the anomaly of having two transfer formats for large scale data, a different format for small scale data and a requirement for the transfer of topologically structured data that caused OS to instigate discussions about a national transfer format.

HISTORY

OS has a number of Consultative Committees, one of which is the Royal Society OS Scientific Committee, which in 1983 had a sub-committee on Digital Cartography (Chairman Professor D W Rhind). A Working Group of this sub-committee considered among other things, the issues involved in defining standards for digital mapping. Some progress was made largely through modelling proposals on the OS transfer standards DMC and OSTF, and in 1984 a trial was carried out to test the transfer of OS digital map data using the British Standards Institution's (BSI) format known as DIAL (Data Interchange at the Application Level). This was not a particularly successful test due mainly to the very different nature of map data from the commercial data for which DIAL was designed. (Smith, 1985).

In 1983 the House of Lords' Select Committee on Science and Technology made the following recommendation (HMSO, 1983):

> "Recommendation 25. Standards for the exchange of digital map data should now be established and consultation to that end should be pressed forward between the British Standards Institution, the OS and other interested bodies, under the aegis of the Royal Society (5.9.2 - 4)."

Following the publication of the Select Committee's report Professor Rhind made a proposal to the Royal Society (RS) that a task force should be established to study the issue of a national transfer standard for digital map data. He proposed that a research student at Birkbeck College, funded through contributions from the main interested Government Agencies, should prepare draft proposals for consideration by the task force.

At the same time the Government Mapping Agencies - OS, Military Survey and the Hydrographic Department - together with the Natural Environment Research Council (NERC) reviewed the situation and it was agreed that the National Working Party under the direction of OS, the main supplier of digital map data, should be established. OS agreed to provide a secretariat and researchers to service the Working Party and an inaugural meeting was held at OS Southampton in February 1985 to consider terms of reference, modus operandi and levels of participation. It was emphasised at the outset that anyone with expertise and a willingness to participate would be welcome, either as a correspondent or a direct collaborator.

Thus although the House of Lords' Select Committee recommended that the national standard should be created under the aegis of the Royal Society, the Government Agencies concerned felt that the most effective way to support the initiative of the Royal Society would be through the creation of a National Working Party led by OS.

THE NATIONAL WORKING PARTY

The National Working Party established two groups - a Steering Group to consider policy, management issues and the implementation of the standard and a Working Group to carry out the work of creating the standard. OS chaired both these groups, provided a research team to support the development of the standard and established the procedures for dissemination of information and the registration of users.

The full title of the working party was "Working Party to Produce "National Standards for the Transfer of Digital Map Data". Until 1989 the steering group was formed from organisations predominantly concerned with the transfer of digital maps and digital data of associated records for display in digital systems.

Members of the Steering Group were drawn from:

Ordnance Survey
Military Survey
Hydrographic Department
Natural Environment Research Council
Royal Society
Royal Institution of Chartered Surveyors
National Joint Utilities Group
Local Authorities Management Services and Computers Committee
Local Authorities Ordnance Survey Committee.

Since 1989 the composition of the Working Party has been widened in an endeavour to reduce the emphasis on digital map data and take account of the increasing need to transfer data for Geographic Information Systems.

Additional members of the Steering Group were invited from:

Science and Engineering Research Council
Economic and Social Research Council
Local Authorities Geographic Information Systems Committee
Association for Geographic Information.

The Working Group was selected from within the organizations mentioned above with the addition of a member from Ordnance Survey of Northern Ireland, and a number of corresponding members were invited to contribute.

Since the establishment of additional working groups to consider specific aspects of NTF, members of commercial organisations involved in hardware manufacture, system design and consultancy have become active in the work of these groups.

TIMETABLE

In order to introduce the new standard as quickly as possible it was agreed that the Auto Carto London Conference in July 1986 would be the target date for the publication of the first draft. (Sowton and Haywood, 1986)

This was achieved and potential users of NTF were invited to consider the draft and make comments. By November 1986 there was sufficient interest and comment to convene a meeting at which criticism and views about NTF could be discussed. The major contribution arising from this meeting was the proposal to simplify the format from a single all embracing standard to one which could take account of various degrees of complexity. As a result the standard was modified and the concept of levels was introduced into the published standard which was issued in January 1987.

The standard has evolved as a result of user comment and the need to resolve difficulties encountered in using the standard both within OS and in the user community.

In January 1989 a complete revision of the standard was issued and Release 1.1 became the current version.

THE PHILOSOPHY

NTF was developed as a transfer format to allow digital mapping data and spatially related data to be passed between users in a simple, but comprehensive manner.

The essential philosophy of the standard was that NTF should be able to conform to three basic requirements:

- NTF should be able to handle all the data generated by any cartographic system. Although GIS was only in its embryonic stage in Great Britain, the likely requirements were considered by the WG. In particular it was essential that forward and back transfer of data should be possible without any discrepancies arising.

- The techniques used for the transfer format must be capable of being adapted easily for any computer on the market.

- The structure of the transfer format should be such that development costs are minimised and maintenance is made easy.

Thus the standard which evolved is both flexible and versatile and with the introduction of Release 1.1 the format has been simplified and the documentation improved.

NTF has been designed to handle a range of data structures and while it may not be particularly suited to every structure, it allows data to be transferred between systems without the need for sender or receiver knowing any details of the other system format. In order to do this without having one complex format, NTF has been split into a number of levels which can be used to transfer

Level 0	Raster and grid data
Level 1	Simple feature coded vector data without topology. A few attributes are possible.
Level 2	As for Level 1 but with the ability to handle an unlimited number of attributes added to the entities.
Level 3	Data which includes topology with polygons and complex objects.
Level 4	Data with an unlimited degree of complication, which can be transferred using this level because it is mandatory to include a data dictionary.

NTF is independent of media, although currently it is more suited to magnetic tape transfer, which remains the dominant medium of transfer. Data can also be transmitted on disc or by telecommunications channel in character form. However, it is necessary to agree between participants the method of transfer, the coding convention and the media and file conventions to be used.

The standard contains sections on the format itself, data quality, glossary and feature coding. (Haywood, 1986)

FEATURE CODING

The first version of the standard included a proposal for a feature classification scheme for both large and small scale topographic data. The recording system proposed was designed to be scale independent and used mnemonic rather than numerical feature codes, thereby relating the code more easily to the feature type being identified.

However this section has not been developed because user opinion appears to support a view that feature coding is a matter for users and a comprehensive listing of all the codes used in UK relating to the transfer of spatially referenced data (even if it were possible for all the various applications) is not considered useful. The approach will be to make feature coding a matter between exchanging groups such as the National Joint Utilities Group (NJUG). The NJUG 11 document (see later) contains a list of feature codes used in transfers between utilities. Similarly topographic data produced by the Ordnance Survey of Northern Ireland and the Ordnance Survey of Great Britain is coded in accordance with the specifications of each organisation.

Consequently this section in the NTF manual is redundant and will be withdrawn.

DATA QUALITY

The great difficulty in defining data quality is to identify a qualitative way in which quality can be sensibly expressed.

The objectives for the assessment of data quality are:

- Information about data quality should cover positional accuracy, attribute accuracy, currency, validity, and completeness.
- The information should be easily accessible; it should be in printed form or within the data being transferred.
- The methods of presenting quality information should be flexible, but they should be clear and understandable; the information should be presented for various levels as appropriate, e.g. at database and feature levels.
- Data quality should be stated as numerical values wherever possible, i.e., it should be quantified to allow comparison with other datasets.
- The information should be scale-independent, e.g., positional accuracy should be stated in terms of ground co-ordinates.

The philosophy behind the proposal is that the user of data should know as much about the history of it as the producer. The implication of this is that producers must make their data self-documenting. Date of survey, method used, algorithms applied, assessment of expected change, the results of accuracy tests, digitising scale, and so on should all be recorded and transferred in digital form. Where appropriate, such information will be applied at feature level, or even to individual co-ordinates.

GLOSSARY OF TERMS

At an early stage in the development of NTF it was realised that if ambiguities in the terminology being used in the documentation were to be avoided, a glossary giving definitions applicable to the transfer of spatially related data and in particular digital map data would be required.

After a great deal of discussion a glossary was accepted which included terms in common use in Great Britain. Terms in use in other countries but infrequently used in Great Britain were excluded, particularly when a more commonly used British equivalent existed.

The Glossary is not a glossary of cartography or a glossary of computing but confines itself to the terms considered necessary to promote a common understanding of the data and the transfer mechanism. These terms

- o define the data being transferred. e.g., Polygon, feature code, raster data.
- o describe the origin or quality of the data. e.g., Edge match, point digitising, accuracy.
- o describe the transfer method or format. e.g., Byte, run length encoding, header label.

The definitions have been based on British usage and not on usage in other countries.

THE FORMAT

The data transfer format was originally designed as two separate parts for raster data and vector data which were later merged into one format. The vector format was subsequently redesigned with the introduction of levels before the standard was published in Version 1.0 in January 1987 and a revised version 1.1 was issued for the vector format alone in January 1989, retaining the 1.0 version for the transfer of raster data. Proposals are now under consideration for a revision of the format relating to the transfer of raster data. While work is in hand to prepare a new version of the vector format which is compatible with ISO 8211.

It is worth considering the objectives which the working group responsible for the early work on NTF established. These remain unchanged and are the basis on which subsequent development will happen.

The underlying reason for a national standard is to improve the efficiency of data transfer, and this can be achieved if three conditions are met:

- o if there is a set of rules for documentation with which every producer and user can become familiar
- o if there is a common method of presentation of data (accepting that some variation will always be necessary)
- o if data can be written and read by the same programs irrespective of its detailed specification.

To achieve this the following objectives were adopted after consideration of earlier work by the US National Committee for Digital Cartographic Data Standards:

- o The format must be as simple as possible.
- o The volume of data transferred in the format should be kept to the minimum required.
- o There should be sufficient internal documentation within the transfer format, to enable the recipient to read the dataset and determine the basic logical structure and physical organisation of the data without having to read external documentation.
- o Transfer formats for vector and raster data should have the same framework, for example, common headers and trailers although the actual formats will differ.
- o The transfer format should allow for the inclusion of all necessary data such as feature information, data quality, spatial and other data types, locational definitions, spatial and other relationships and ancillary data.
- o The standard should include a data dictionary.
- o Industry standards, where applicable, should be adopted to handle the various types of data.
- o The transfer format should be media-independent.
- o The transfer format should be computer-independent.
- o Entities and concepts must be transferrable and the format must be flexible enough to allow for the transfer of new concepts without modifying existing ones.
- o There must be the capability for transferring "change-only" information to update databases.
- o The format should allow for the transfer of logically different datasets on one physical medium.

THE INTRODUCTION OF LEVELS

The introduction of levels has made the use of NTF simpler for the transfer of simple feature coded vector data without topology but which may include attribute data. The use of levels has not made a great change in the handling of topologically structured data, although level 3 is less complex than level 4.

VERSION 1.1

Version 1.1 was published in January 1989 after a complete review of the standard. The original concept that data could migrate from lower levels to higher levels as it became more sophisticated was found not to work and also to be an unnecessary complication between the levels.

In version 1.1 the distinction between the vector levels of transfer was clarified and the design parameters were re-assessed. Given that common containing principles and structures (the framework) should apply whatever type of spatial data is being transferred it was seen that the important features of NTF are:

- Users of a lower level such as Level 1, do not need to know how the format applies at higher levels once the decision to adopt a particular level has been made.

- There need not be any similarity between records used in the different levels. The reason for changing from one level to another would be a perceived advantage in using different or more sophisticated data rather than the ease in which the change to a more complex transfer format can be carried out. However, if similarity can be achieved between the data being transferred in the various levels, this would have benefits in simplifying such a change but is not a necessity.

- The degree of flexibility does not have to be the same at all levels. The most flexible level is also the most complex and thus a balance has to be struck between the cost and the desirability of moving to Level 4 from one of the lower levels.

CURRENT STATUS

NTF has become an accepted standard format for the transfer of digital map data between UK organisations. Northern Ireland is currently using the standard and is participating in the Working Group. In addition there are a number of other countries including the Republic of Ireland and Hungary who have adopted NTF or are using a variant of it for the transfer of digital map data. The GDF (Geographic Data

Files) standard created by Philips and Bosch for the transfer of in-car navigation data is based on NTF. (DEMETER, 1988)

The OS support for NTF has ensured that suppliers of systems have provided NTF interfaces to allow interchange using NTF to happen between differing systems.

The National Joint Utilities Group has recently issued a new proposal for the transfer of spatially related data between utilities using NTF. Their publication number 11 gives a comprehensive description of NTF and the way in which it will be used to transfer utility data. (NJUG 1990).

ISO 8211

In June 1990 a review of NTF was undertaken with a view to making it compatible with ISO 8211.

It has been possible to use functions provided by ISO 8211 to simplify special features provided in the levels of NTF. As a result common data descriptions have been used for levels 1 and 2 whereby level 1 becomes a subset of level 2. Similarly it has been possible to create a common data description for levels 3 and 4. Although it would be possible to merge all the levels it was considered that this would complicate the format unnecessarily and would remove one of the important features of NTF which is to keep lower level transfers simple. One further important advantage will be the opportunity to add new fields into level 4 which will allow spatial reference systems which are not coordinate based to be introduced without upsetting the normal operation of the format.

Work is now in hand to complete the documentation of the changes and to perform some simple experiments with the format. Once this has been done the members of the Working Group being set up to evaluate these changes will be asked to consider how effective the new format is, and make recommendations before the Steering Committee considers whether to adopt the changed format.

In parallel with this work the Standing Technical Working Group will be reviewing the documentation and will consider what additional improvements not taken into account by the ISO 8211 changes are necessary.

It is expected that during 1991 the NTF vector formats will have been comprehensively revised, software for the ISO 8211 implementation will be available and NTF will be wholly compatible with ISO 8211.

CONTROLLING ORGANISATIONS

During 1989 negotiations were carried out between the National Working Party and the newly formed Association for Geographic Information (AGI) whereby the responsibility for

NTF could be transferred to the Standards Committee of AGI. Agreement was finally reached and AGI is now the controlling body for NTF.

While the change has not affected the way in which NTF is managed technically it has improved the business and administrative arrangements which are now completely independent of Ordnance Survey.

The NTF Steering Group has been expanded and new terms of reference have been written together with a change in the name of the group which is now the NTF Steering Committee. The terms of reference are as follows:

- o To develop NTF as a suitable transfer mechanism for digital map data and map related data compatible with the Standards which evolve for GIS.
- o To encourage the use of NTF in UK and Europe.
- o To maintain the standard in line with the user community.
- o To ensure compatibility with other transfer standards, eg ISO, CAD/CAM.
- o To liaise with other foreign agencies developing standards, eg CERCO, ICA, DGIWG, AM/FM (Europe), IHO.
- o To issue amendments.
- o To keep users informed through a newsletter.
- o To create and monitor a business plan.
- o To create or dissolve working groups as necessary, to determine their terms of reference and to monitor their progress.
- o To report to the Standards Committee of the Association for Geographic Information.

Three more Working Groups have been created and there are now:

- o The Standing Technical Working Group responsible for maintenance and revision of the standard - Chairman Mr. P. E. Haywood.
- o The ISO 8211 Working Group responsible for overseeing a consultancy to make the standard ISO 8211 compatible - Chairman Mr. T. Ibbs.
- o The Business Strategy Working Group responsible for the budget and financial considerations - Chairman Mr. M. Sowton.

- The Level 0 Working Group responsible for reviewing and improving the handling of raster data - Chairman Mr. A. R. Rutherford.

CONTACT

NTF is now administered through the AGI who are responsible for the secretariat and register of users of NTF. Currently new members pay a fee for the initial documentation and thereafter an annual registration fee which entitles participants to newsletters and amendments to the standard.

Details of the membership arrangements and copies of the NTF documentation can be obtained from

The Secretary
Association for Geographic Information
Royal Institution of Chartered Surveyors
12 Great George Street
Parliament Square
London SW1 3AD

Correspondence about technical matters relating to NTF particularly suggestions for change or improvement, should be sent to the Secretary of AGI who will establish contact with the most appropriate part of the NTF Working Party.

CONCLUSION

NTF is now administered and controlled by AGI and is independent of major producers of data and systems although representatives of these organisations participate in the Working Party.

It is expected that an ISO 8211 compatible format for vector data will be introduced towards the middle of 1991, but improved raster handling will be delayed until 1991.

The format will be reviewed and revised documentation will be produced initially for vector data and for raster data as soon as possible thereafter.

NTF is an established working format which has evolved through use into the current version 1.1. However any registered user is welcome to make suggestions for improving the format. Such suggestions are reviewed at intervals by the appropriate Working Group and amendments are promulgated to users through the Newsletter. The evolution into the ISO 8211 version will be treated in this way and a re-written NTF manual will be issued to members after acceptance by the Steering Committee.

REFERENCES

DEMETER, 1988, "Geographic Data Files (GDF) Release 1.0,"

EUREKA Project - Digital Electronic Mapping of European Territory, (available from Divisional Standardisation Department of Philips Consumer Electronics, Eindhoven, Contact DEMETER Secretary).

Haywood, P., 1986, "National Standards for Great Britain." In Land and Minerals Surveying, Vol 4, No. 11, ed. M. Schwartz RICS, London, pp 569 - 578.

Her Majesty's Stationery Office, 1983, "Report of the Select Committee on Science and Technology - Remote Sensing and Digital Mapping," chaired by Lord Shackleton, HMSO, London.

International Standards Organisation, 1985, "ISO 8211 - Information Processing - Specification for a data descriptive file for information interchange".

National Joint Utilities Group, 1990, "Proposed Data Exchange Format for Utility Map Data", Publication No. 11, Issue No. 2, NJUG, London.

Smith, N. S., 1985, "The UK Experience in Interchanging Digital Topographic Information", Proceedings Survey and Mapping Conference, Reading, 1985 Paper G4, 9 pages.

Sowton, M. and Haywood, P., 1985, "National Standards for the Transfer of Digital Map Data", Proceedings of Auto Carto - London, ed. M. Blakemore, pp 298 - 311.

THE UNITED STATES SPATIAL DATA TRANSFER STANDARD

Hedy J. Rossmeissl
U.S. Geological Survey
561 National Center
Reston, Virginia 22092
U.S.A.

Robert D. Rugg
Virginia Commonwealth University
Box 2008
Richmond, Virginia 23284
U.S.A.

ABSTRACT

Significant developments have taken place around the world in the disciplines of cartography and geography in recent years with the advent of computer hardware and software that manipulate and process digital cartographic and geographic data more efficiently. The availability of inexpensive and powerful software and hardware systems offers the capability of displaying and analyzing spatial data to a growing number of users. As a result, developing and using existing digital cartographic data bases is becoming very popular. In the United States, the absence of uniform standards for the transfer of digital spatial data is hindering the exchange of data and increasing costs.

In the United States, both the Federal Government and the academic community have been working hard over the last few years to develop the Spatial Data Transfer Standard that includes definitions of standard terminology, a spatial data transfer specification, recommendations on reporting digital cartographic data quality, and standard topographic and hydrographic entity terms and definitions. This proposed standard was published in the January 1988 issue of The American Cartographer. Efforts to test and promote the Spatial Data Transfer Standard were coordinated by the U.S. Geological Survey. A Technical Review Board was appointed with representatives from the U.S. Government, the private sector, and the university community to finalize the standard and submit it to the U.S. National Institute of Standards and

Technology for approval as a Federal Information Processing Standard. The Standard was submitted to the National Institute of Standards and Technology in July 1990.

NECESSITY OF DATA INTERCHANGE STANDARDS

Early in the 1980's geographers and cartographers in the United States recognized that the need to transfer spatial data between noncommunicating systems was becoming increasingly important. The concerns for common data formats and geocoding conventions cut across all topics of spatial data handling. It was difficult and inefficient for diverse users to use a given set of data. At least five major forces were causing concern about incompatibility: (1) accumulation of increasing amounts of spatial data to be stored, cataloged, and retrieved; (2) rapid progress and expansion in the area of spatial data processing; (3) expansion in the amount of related and useful data being collected in digital form; (4) collection of more detailed digital data because of increasing sophistication in the ability to register digital images with maps, as well as the analysis of multiple sets of data; and (5) perpetuation of duplicative and redundant map automation efforts. (Digital Cartographic Data Standards (DCDS) Task Force, 1988)

The U. S. Spatial Data Transfer Standard (SDTS) was developed to offer to users of spatial information: (1) standard data models and terminology, (2) a systematic and comprehensive set of primitive and simple cartographic objects from which digital cartographic feature representations can be built, (3) the transfer of digital spatial information between noncompatible systems while preserving the meaning of the information, (4) data quality information in order to evaluate the fitness of data for a particular use, (5) standard topographic and hydrographic entity terms and definitions, and (6) a physical file format specification to implement the standard. Having such a comprehensive standard in place will be a great benefit to the users of digital spatial data in the United States.

HISTORY

In 1980 the U.S. National Bureau of Standards (renamed the U.S. National Institute of Standards and Technology (NIST)) signed a memorandum of understanding with the U.S. Geological Survey (USGS) that resulted in the USGS assuming the leadership in developing, defining, and maintaining earth science data elements and their representation standards for use by U.S. Government agencies.

Beginning in 1982 a national committee in the United States, composed of members from private industry, government, and academia, systematically developed a Digital Cartographic Data Standard. The National Committee for Digital Cartographic Data Standards was formed under the

American Congress on Surveying and Mapping and supported by a grant from the USGS. Professor Harold Moellering of The Ohio State University headed the national committee, which produced nine reports. The work of the committee culminated in Report 8, "A Draft Proposed Standard for Digital Cartographic Data" issued in 1986.

The Office of Management and Budget issued a memorandum in 1983 that directed the USGS to eliminate duplication and waste in the development of Federal digital cartographic data bases and to serve as a focal point for coordination of digital cartographic activities. In response to this memorandum, the USGS established the Federal Interagency Coordinating Committee on Digital Cartography (FICCDC). A standards working group under the FICCDC selected as its priority task the development of a data exchange format and published its report in 1986 entitled "Federal Geographic Exchange Format."

The USGS established a Digital Cartographic Data Standards Task Force in March 1987 to meld the proposed cartographic data standard with the geographic exchange format. This combined document was published as the January 1988 issue of *The American Cartographer*.

After the publication of *The American Cartographer* the responsibility for finalizing the standard reverted to the USGS. A maintenance authority within the structure of the USGS National Mapping Division was established. The tasks of the maintenance authority included testing the standard, conducting educational workshops, disseminating information, and coordinating all activities surrounding the promotion of the standard to NIST as a Federal Information Processing Standard (FIPS). Dr. Joel Morrison of the USGS chairs the Technical Review Board that oversaw changes to the content of the standard as it was being prepared for submission to NIST. The experts serving on the Board believe strongly in the benefits of standards and yet represent a diversity of viewpoints in the collection and use of digital spatial data. They have worked together to generate thought-provoking discussions and well thought-out recommendations.

OBJECTIVES

The objectives of the standard as stated by the DCDS Task Force (1988) are:

- To provide a mechanism for the transfer of digital spatial information between noncommunicating parties by using dissimilar computer systems, by preserving the meaning of the information, and by reducing to a minimum the need for information external to this standard concerning the transfer.

- To provide, for the purpose of transfer, a set of clearly specified spatial objects and relationships that represent real world spatial entities and to specify the ancillary information that may be necessary to accomplish the transfers required by the cartographic community.

- o To provide a transfer model that will facilitate the conversion of user-oriented objects, relationships, and information into the set of objects, relationships, and information specified by this standard for the purposes of transfer in such a way that their meaning will be preserved and can be discerned by the recipient of a conforming transfer.

- o To ensure that the implementation of this standard has the following characteristics:

  - the ability to transfer vector, raster, grid, and attribute data and other ancillary information;

  - the implementation methodology can be media independent and extendable to encompass new spatial information as needed;

  - an internally contained description of the data types, formats, and data structures in such a way that the information items can be readily identified and processed into the user's native system; and

  - the data and media formats should be based where practical on existing FIPS, American National Standards Institute (ANSI), International Standards Organization (ISO), or other accepted standards.

CONTENT

The SDTS provides a comprehensive solution to the problem of spatial (that is, geographic and cartographic) data transfer, from the conceptual level to the details of physical file encoding. Transfer of spatial data involves modeling spatial data concepts, data structures, and logical and physical file structures. To be useful, the data to be transferred must also be meaningful and adequate in terms of data content and data quality. The SDTS addresses all of these aspects of data transfer for both vector and raster data structures.

The standard is in three parts. Part 1 addresses the logical specifications in terms of conformance requirements, a conceptual model, quality specifications, the data structure model, and the transfer format. Part 2 addresses data content by providing a standard list and definitions of spatial features and their attributes. Part 3 specifies the implementation of the SDTS in terms of the ISO standard for a Data Descriptive File for Information Interchange.

Section 1 of part 1 includes a statement of scope and conformance requirements for the SDTS. It also includes normative references to other standards and definitions of terms.

Conformance to the SDTS is viewed in two ways. First, there are various kinds of users with different "profiles." The user profiles are identified in the context of the SDTS model of spatial data (see below). Second, there are specified levels of conformance for spatial addressing systems and for spatial features. For example, there are three levels of conformance to spatial addressing:

Level 1 spatial addresses appear as geographic coordinates (that is, latitude and longitude) or one of the two alternative legally recognized systems in the United States (that is, Universal Transverse Mercator or State Plane Coordinate Systems) that have a known mathematical relationship to latitude and longitude.

Level 2 spatial addresses are given in some other projection system that has a known relationship to latitude and longitude. Included in the transfer are the name of the projection system and the translation parameters for conversion to latitude and longitude.

Level 3 spatial addresses have no known relationship to latitude and longitude. This option is most likely to be used for certain kinds of raster data or for detailed graphic representations of large-scale features.

The conceptual model of spatial data is presented in section 2 of Part 1 to provide a framework for defining spatial features and a context for the definition of a set of spatial objects. This conceptual model supports the translation of user data models to and from the SDTS model. Within section 2 is a defined set of spatial objects for zero, one, and two dimensions used in spatial data systems to represent real-world spatial phenomena. Three-dimensional spatial objects have not been specified. The defined set of objects will support the three major kinds of spatial data operations: (1) geometry-only, (2) geometry and topology, and (3) topology-only. These objects have been specified in a modular fashion in such a way that more elaborate composite objects can be constructed from them (see Table 1).

Table 1. Defined Primitive Spatial Objects in Zero, One and Two Dimensions

Dimension	Geometry only objects	Geometry and Topology objects
zero	point	node
one	string arc line segment G-ring	link chain GT-ring
two	G-polygon grid cell pixel	GT-polygon

The conformance profiles of users recognize these differences. It is not anticipated that the SDTS will solve the problem of direct translation of raster data in one system to vector data in another system or conversion of geometry-only data in one system to topologically structured data in another. Rather, the transfer will include the information as to what kinds of data are included. The kinds of spatial data profiles have been categorized as follows:

Geometry only vectors (points, lines, and areas);

Vector data with geometry and topological relationships identified;

Raster data based on grid cells or pixels; and

Data including "composites," that is spatial objects derived from assembling primitive objects of one or more of the above types into a more complex representation of a particular real world entity.

The profile of a given transfer specifies which of these four kinds of spatial data are present and which are absent.

Section 3 of Part 1 includes specifications for a quality report concerning the objects in a transfer and their attributes. The purpose of the quality report is to provide detailed information for a user to evaluate the fitness of the data for a particular use. This style of standard can be characterized as "truth in labeling," rather than fixing arbitrary numerical thresholds of quality. To implement this portion of the standard, a producer is urged to include the most rigorous and quantitative information available on the components of data quality described in this section (see Table 2). This kind of information is essential for the intelligent use of spatial data, and yet most of it is generally absent from many currently supplied spatial data products.

Table 2. Components of Spatial Data Quality

Component	Definition and Examples
Lineage	Description of source material used and the methods of derivation, including mathematica transformations of coordinates.
Positional Accuracy	Degree of compliance to the spatial address standard, including information on control surveys and the accuracy of spatial address in the final product determined by deductive estimate, internal evidence, comparison to source, and (or) tests using independent sources of higher accuracy.
Attribute Accuracy	Deviations of measures of attribute values from the true values based on tests similar to those used for positional accuracy, possibly including a misclassification matri of sample points or polygon overlay areas fo nominal scale attributes.
Logical Consistency	Fidelity of relationships encoded in the dat structure, including tests of valid values, general graphic tests, and (or) tests for topological consistency.
Completeness	Information about selection criteria, definitions used and other relevant mapping rules such as minimum area thresholds, use c standard geocodes, and exhaustiveness of areal coverage.

Sections 4 and 5 of Part 1 present specifications for the transfer of spatial data. Logical modules consisting of detaile record, field, and subfield specifications are presented in section 5. Section 4 contains general concepts and specifications and the underlying models that pertain to the transfer module specifications of section 5. Section 4 also specifies the general elements of an implementation, the relationships of the logical constructs of the data models to th general elements of a detailed implementation, and general constraints on the implementation. Finally, section 4 presents the notational conventions used in the section 5 module specifications.

Part 2 of the standard responds to the need for standardizing spatial features. The conceptual model and the defined terms are the foundation for a uniform approach to creating a spatial feature file and to exchanging data between existing digital cartographic systems. Given the widespread absence of definitions in digital cartographic data, the standard requires that complete definitions of entity and attribute terms be supplied along with the data in a transfer. Part 2 provides a standard set of terms and definitions that should be used unless additional terms or non standard usage of terms appears in the data.

The model of spatial features used for part 2 is a non hierarchical classification of real world entities into entity types such as airport, boundary, forest, island, railway, tower, and watercourse. These entity types may be generalized by users of hierarchical systems into themes, but they may also be incorporated in more complex data structures such as networks and object-oriented systems.

The standard provides 200 initial entity type definitions drawn from hydrographic and topographic mapping applications. The standard definitions of these 200 standard entity types represent the distillation of 2,400 definitions of over 1,200 entity terms previously used by various government agencies to describe essentially similar real world phenomena. The standard retains the list of 1,200 "included terms" and shows which standard term should be used to represent it in a standard transfer.

Along with standard entities, part 2 includes an initial set of 245 standard attributes and some additional "included" attribute terms. In principle, any standard attribute may be used with any standard entity, which provides flexibility for users to adapt the standard terms with standard meanings to a variety of possible attribute data structures.

There are two intended applications of this part of the standard for systems employing spatial data in digital form: (1) new users will have a model and the definitions needed for creating a features list, and (2) the transfer between existing systems will be facilitated. Successful transfer implies not only the receipt of a file and placement of the data in the appropriate fields in a receiving system, but also proper semantic interpretation of the data received. This part of the standard requires that users attempt to translate their own entity definitions to the combination of standard entity and attribute terms that will ensure a commonly understood definition of the real world phenomena about which data are being transferred. If this cannot be done, users must supply complete definitions for the non standard terms or codes appearing in their data. Table 3 shows the various levels of conformance to the spatial features portion of the standard.

Part 3, the ISO 8211 encoding, provides a representation of a spatial data transfer file set. This encoding was selected with the following objectives:

- o to provide the syntax and semantics necessary for transporting files, records, fields, and subfields accompanied by their data description in machine-readable form;

- o to provide for media independence of the file set; and
- o to provide for compatibility with this approved national an international general purpose standard for data exchange.

Table 3. Conformance Levels for Spatial Features

level 1	Only standard entity and attribute terms and definitions are used in the transfer
level 2	Non-standard entity and (or) attribute terms are used, but they have been systematically compared with the standard lists of "include terms" and mapped where possible to standard terms. The remaining non-standard terms hav been fully defined in an accompanying data dictionary module.
level 3	Some non-standard entity and (or) attributes terms are used, and they have not necessaril been checked against the standard lists of included terms. However, full definitions are given for the non-standard terms in a data dictionary module.
level 4	Only non-standard entity and attribute terms appear in the transfer. Each term or code i fully defined in a data dictionary module.

As time goes on, this part of the standard will be revised and updated to incorporate additional entity types to be transferred among user organizations.

STATUS

The proposed Digital Cartographic Data Standard (DCDS Task Force 1988) was tested in two phases between January 1988 and April 1989. During phase 1 testing agencies most involved in the development of the standard assembled test data sets. The USGS encoded a Digital Line Graph (DLG) file, the Bureau of the Census, a TIGER file, and the Defense Mapping Agency, a world vector shoreline file. These files were exchanged and decoded b each participating agency. Errors and suggested improvements resulting from this exercise were documented and presented to th Technical Review Board for review. The Board voted to accept 14 of 17 proposed changes. These changes were incorporated in the standard and a revised version "The Proposed Standard for Digita Cartographic Data" (DCDS Task Force, 1988) was prepared for the phase 2 testing period.

Additional support documentation, the Digital Cartographic Data Standards Procedure Manual (U.S. Geological Survey, 1988), was prepared for phase 2 testers. This manual is a companion document to the proposed standard and offers a summary of the standard, a description of testing procedures, and guidelines for using the standard.

Phase 2 testing began in September 1988. Three workshops were held to inform participants of the background, purpose, goals, and content of the standard. Over 60 individuals from more than 20 U.S. Government agencies, private firms, and universities attended. Participants were asked to test the concepts presented in the standard and report their finding to the Standard Maintenance Authority. As a result of the testing 124 suggestions were submitted. Major concerns were related to the number of options in, and the complexity of, the standard, and suggestions were made to clarify specific points.

The Technical Review Board reviewed all comments and completed a substantial amount of work to refine the standard including defining a data model, restructuring the vector and raster transfer modules, providing a flexible approach in designating map projections and a coordinate system, specifying user profiles, and simplifying the documentation. The Board recognized that completing the changes mandated through testing would cause a delay in the submission of the standard to NIST, however a thorough rework was deemed necessary and worthwhile. The changes being proposed would simplify the documentation, decoding tasks, and attribute encoding. These measures retained the properties necessary for a robust standard that handles both vector and raster data but also served to create a more easily understandable and usable document.

Upon completion of this rework a period of thorough editorial review was undertaken to identify and correct ambiguities, discontinuities, and misstatements. This review was completed by those individuals and organizations that participated in the previous testing periods. Upon completion of this review the SDTS was finalized and submitted to the NIST (July 1990).

The NIST provided technical assistance to support the final phase of preparing the proposed SDTS for FIPS processing. Support was provided to evaluate the SDTS in terms of FIPS standards and guidelines for data administration, data base management, graphics, and related standards; to prepare a strategy for successful standardization of the SDTS including the identification of potential problems and issues with recommendations for resolving them; and to initiate the planning, development, and implementation of a conformance testing program for the SDTS. This coordination was undertaken to ensure that potential problems could be identified before the standard was finalized and to avoid extended delays and rework during the FIPS approval process.

The SDTS is currently in the early stages of this approval process. The maintenance authority, located in the Office of Technical Management of the USGS National Mapping Division, is turning its attention to responding to comments arising during the FIPS process, to public information about the standard, and to the development of improved user support capabilities. In addition, the SDTS staff will work to help implement the developing National Geographic Data Base System described below.

The USGS is prepared to submit this standard to the ANSI fo promotion as an ANSI standard upon its approval as a FIPS. An internationalized version may be proposed as an ISO standard.

UTILITY

User Interface
Users are a vital part of this development effort. The latest improvements to the standard were initiated in part because of concerns about the difficulty of using and promoting a complex standard. The USGS recognizes that a significant effort must be undertaken to develop software interface tools and documentation so that the detailed constructs contained in the standard will b transparent to the user.

The vendor community must also take an active role in developing software tools to encode and decode spatial data between the standard format and their own system format. A concerted effort has been made during the development of the standard to keep the private sector informed and to use their suggestions to improve the standard. With the cooperation and support of many Federal agencies and the private sector, the transition of converting to this standard can be made smoothly and effectively.

National Geographic Data Base System
Within the U.S. Government developing and using existing digital cartographic data bases is becoming very popular. The absence o uniform standards for the transfer of digital spatial data is hindering the exchange of data and increasing costs. From the Federal agency perspective the use of the SDTS is enormous. Wit such a standard in place a large distributed spatial data base system within the Federal Government is possible. The National Geographic Data Base System would be a system of independently held and maintained Federal digital spatial data bases. Traditional cartographic categories such as hydrography, boundaries, transportation, and elevations would be held, with the addition of thematic categories such as soils, wetlands, geology, vegetation, and demography. The objectives of this system are to:

- encourage and hasten the use of the SDTS;
- provide a mechanism for improved coordination and standardization of data content and quality;
- provide for more consistent and compatible Federal spatial data bases;
- improve the efficiency and effectiveness of Federal agencie and others using Federal spatial data bases;

- o inform users and potential users of digital spatial data regarding availability and content; and

- o clarify Federal agency responsibilities for developing, maintaining, and distributing spatial data bases (Federal Interagency Coordinating Committee on Digital Cartography-Standards Working Group, 1990).

Such a system would require each agency responsible for spatial information about the United States to be able to release its information in the SDTS format. The updating and accuracy of a data set would remain the responsibility of the agency that has the mandated responsibility to collect and use that data set. The National Mapping Division of the USGS will establish our base data categories, such as hypsography, hydrography, transportation, and boundaries, etc. in data files that will be accessible by the general public through the SDTS. It is envisioned that other agencies such as the Soil Conservation Service will have basic soils data available in a similar manner so that a user could access soils data and overlay it on USGS base data without knowledge of the internal formats used by each agency. This allows each agency to design and use its own internal format to process spatial data and does not require standardization across the Federal Government on one data structure, data format, and hardware/software system.

CONCLUSION

The time for a spatial data standard is here. The use of standards has many advantages to the data collector, processor, and user, particularly those who need data from several sources. The availability of standard terminology will be of great benefit in setting a common language of definitions for users and producers of spatial data. A standard transfer specification will facilitate the exchange of spatial data throughout the public and private communities by offering the capability of displaying, analyzing, and integrating spatial data for a growing number of applications. The availability of quality information will provide users details about the data including lineage, completeness, accuracy, and logical consistency in order to evaluate the fitness of the data for a particular use. The SDTS provides these characteristics and offers a mechanism for the exchange of spatial data that is long overdue.

REFERENCES

American National Standards Institute, Inc., 1986, Specification for a Data Descriptive File for Information Interchange (ANSI/ISO 8211-1985), FIPS PUB 123.

American National Standards Institute, Inc., 1986, Computer Graphics - Metafile for the Storage and Transfer of Picture Description Information. ANSI X3.122-1986, FIPS PUB 128, 289 pp. plus annexes.

Digital Cartographic Data Standards Task Force, 1988, The Proposed Standard for Digital Cartographic Data, dedicated issue, The American Cartographer, 15:1.

Federal Interagency Coordinating Committee on Digital Cartography--Standards Working Group, 1990, "Guidelines and Benefits to Participants of the National Geographic Databases System," (available from U.S. Geological Survey, Office of Geographic and Cartographic Research, Reston, VA 22092, USA).

Moellering, Harold (ed.), 1987, Issues in Digital Cartographic Data Standards, published by the U.S. National Committee for Digital Cartographic Data Standards, Report #9, "A Final bibliography for Digital Cartographic data Standards".

Morrison, J.L., 1988, Digital Cartographic Standards: The United States Experience. Proceedings of 7th Australian Cartographic Conference (Austro Carto III), Sydney, Australia,).

Spatial Data Transfer Standard Technical Review Board, 1990, Spatial Data Transfer Standard, (available from U.S. Geological Survey, Office of Technical Management, 510 National Center, Reston, VA 22092.)

U.S. Geological Survey, The Digital Geographic Data Standard Procedure Manual, 1988 (available from the U.S. Geological Survey, Office of Technical Management, 510 National Center, Reston, VA 22092.)

CERCO CONSIDERATIONS CONCERNING AN EUROPEAN TRANSFER FORMAT

M. SOWTON
Member of CERCO Working Group V
Ordnance Survey, Rosmey Road
Maybush, Southampton, UK

ABSTRACT

This chapter describes the progress towards the creation of an European Territorial Database (ETDB) and the associated considerations of a suitable European Transfer Format compatible with both the ETDB data model and ISO 8211 (ISO, 1985). Brief details about an exchange format for navigation data are included.

INTRODUCTION

CERCO (Comité Europeen des Responsables de la Cartographie Officielle) is a committee at which the Heads of the national agencies responsible for official cartography in Western Europe meet to consider issues of mutual concern. CERCO has a number of Working Groups to consider the details of such issues, two of which are concerned with data standards: WG V and WG VII. (CERCO, 1988)

WG V is involved with the definition of an European Territorial Database (ETDB) which aims to contain the minimum content of each national database. This leads to the definition of an exchange standard for geographical data.

WG VII is involved with the definition of a topographic database for road navigation systems. It is collaborating with two programmes - the EUREKA sponsored PROMETHEUS programme and the European Community funded DRIVE programme. The EUREKA programme is an initiative at ministerial level created by a group of European nations (not only from the European Community) to encourage collaboration on research and development projects with the aim of increasing the European competitive base in technology. PROMETHEUS (Programme for an European Traffic with Highest Efficiency and Unprecedented Safety) was established as a EUREKA programme by the European motor industry with the intention of improving all aspects of road transport. One aspect of this programme is concerned with road navigation and the management of road traffic through guidance, route planning and the provision of information to drivers. DRIVE (Dedicated Road Infrastructure for Vehicle Safety in Europe) is an European Community programme of collaborative research and development which seeks to alleviate some of the present problems in road transportation such as road safety, traffic efficiency and pollution.

WORKING GROUP V

WG V of CERCO held its inaugural meeting in Brussels from 5th to 7th November 1986 under the Chairmanship of Colonel Rosario Talamo of IGM - Florence.

The purpose of the group was to create an ETDB. The following terms of reference were adopted at the first meeting:

- to create a standardized model of the information contents for digital mapping at small, medium and large scales and their relevant symbology.
- to elaborate a digitized data structure model as well as the corresponding exchange format.
- the activities of Working Group V also aim at supporting the cartographic requirements of the CORINE Project, as decided by CERCO during its Plenary Assembly in Belfast in September 1986. CORINE (Coordinated Information on the European Environment) is an European Community programme to gather consistent information on the resources and characteristics of the environment most directly affected by development programmes of which information about land use is a high priority. The ETDB is intended to support the display of the CORINE results.
- membership of WG V is open to all CERCO member countries.

SUB-GROUPS

At the first meeting it was decided that two sub-groups should be formed in order to speed up the work. Separate meetings were held in Florence and Bonn.

- Florence (Italy, Belgium, Spain) — Minimum informative contents
Coding System
Database Management System

- Bonn (Germany, Denmark, France) — Data structure, Exchange format

At the second meeting of the Working Group, as a whole, held in Florence the work of these two sub-groups was considered and proposals about the informative contents, coding system and data structure were agreed for presentation to the Plenary Assembly of CERCO in Athens in 1987.

It was considered that it was too early to make proposals about the exchange format for the Athens meeting because the problem of creating a format which would allow exchanges between databases with different structures had not been studied. However it was decided to produce a report about the possible alternative data structures which could be used.

OPTIONS FOR A TRANSFER FORMAT

At the 3rd meeting held in Madrid from 4th to 7th May 1988 Mr. Beltrame of the National Research Council in Pisa, Italy gave a comprehensive presentation about the FGEF exchange format and ISO 8211 and Mr. Sowton presented an outline of the NTF format in use in Great Britain (NTF, 1989). Discussion at this meeting centered on the complication of the ISO standard, the fact that NTF was actually being used (FGEF was at that time only at the trial testing stage) and the possibility of NATO adopting ISO 8211 as a NATO standard on the recommendation of the Digital Geographic Information Working Group (DGIWG). It was agreed that three alternatives would be considered.

- FGEF
- NTF
- ISO 8211

EUROPEAN TRANSFER FORMAT

The 4th meeting of WG V took place in Paris between the 31st January and 3rd February 1989. This was the first time that a definition of the requirements of the transfer format were stated.

- to make possible the physical exchange of data
- to enable the geographic content of the data to be understood

It was made clear that the creation of an ETF would not impose an exchange mechanism on CERCO members but rather the creation of ETF would offer each separate country the alternative of:

o writing an interface from their national format into ETF

or

o adopting ETF.

The Paris meeting adopted a two stage approach first an interim stage would be reached when as a temporary measure NTF, GDF or some other alternative would be used, followed by the adoption of an ISO 8211 compatible standard. It was agreed that if NTF could be made compatible with ISO 8211 and could accept the adopted data model and structure of ETDB it should become the first basis for an interim ETF.

The report of the Paris meeting included recommendations to:

o prepare an initial draft of ETF

o propose recommendations about the maintenance of the standards and ETF.

The decision to consider NTF as a main candidate for ETF was taken because at that time it was the only available exchange format which actually had the format defined and in use, and which was expected to be adaptable with the least effort to the data structure proposed for ETDB. However a proviso was made that NTF had to be made compatible with ISO 8211 before it could be accepted as the ETF as well as being compatible with the structure being adopted for the ETDB.

This decision was confirmed at the 5th meeting of the WG V held in Ankara from 31st January until 2nd February 1990, although there was a counter proposal that the DIGEST format should also be considered. Details about DIGEST are given elsewhere in this Monograph.

NAVIGATION DATA

A EUREKA project called DEMETER (Digital Electronic Mapping of European Territory) has established GDF (Geographic Data Files) as a format for the handling of navigation data. This format is also the internal format of navigation systems developed by Philips and Bosch. GDF has been accepted as a transfer format for navigation systems and CERCO WG VII which is concerned with the definition of a topographic database for road navigation systems has also adopted this format. Notwithstanding the fact that GDF is based on NTF, digital mapping data and spatial data are more complicated and have a greater content, consequently WG V could not adopt this standard. Thus an interface between mapping formats and GDF will be required.

CURRENT STATUS OF NTF

Dr. Alfred A Brooks has recently carried out a consultancy with a view to making NTF compatible with ISO 8211. Dr. Brooks who was leader of the team responsible for creating ISO 8211 came to Britain at the end of June 1990 to carry out the work. Tests are now being carried out by an NTF Working Group and at the end of this work an ISO 8211 version of NTF will be available for consideration as a candidate for ETF.

Although no approach has yet been made to the British Standards Institution (BSI) to have NTF adopted as a British Standard, NTF is now a de facto standard in UK for the transfer of digital mapping data and spatially related data. A number of system suppliers have written NTF interfaces. Responsibility for NTF has recently been transferred to the Association for Geographic Information (AGI) who will manage and maintain NTF until such time as an approach to BSI is considered necessary. The benefit of having AGI involved with the management and technical direction of NTF is that there is now a sound financial base and there is no single interest group overseeing the standard.

A more detailed description of NTF is given elsewhere in this Monograph.

CONCLUSION

Until the outcome of the ISO 8211 compatibility is known and has been considered by CERCO, it is not possible to determine whether NTF will become accepted as the European Transfer Format.

Once this conversion has been carried out trials will be performed to determine the suitability of the result as an ETF.

It will be possible to evaluate alternative candidates such as DIGEST at this time.

GDF is not a suitable transfer format for all spatial data but is highly suitable as a format for navigation data and associated attributes.

Although recommended at the Paris meeting arrangements for the maintenance and management of an European Transfer Format have still to be considered.

REFERENCES

CERCO, 1988, "The European Committee of the Heads of the Official Mapping Agencies," 12 pages (available from The Secretary General - CERCO Mr. J. Mousset, IGN, Brussels).

DEMETER, 1988, "Geographic Data Files (GDF) Release 1.0," EUREKA Project - Digital Electronic Mapping of European Territory, (available from Divisional Standardisation Department of Philips Consumer Electronics, Eindhoven, Contact DEMETER Secretary).

International Standards Organisation, 1985, "ISO 8211 - Information Processing - Specification for a data descriptive file for information interchange".

NTF, 1989, "The National Transfer Format Version 1.1", (available from the Secretary Association for Geographic Information, RICS, London).

DIGITAL GEOGRAPHIC INFORMATION WORKING GROUP: EXCHANGE STANDARDS

Ian Smith
Systems and Techniques Unit
Directorate General of Military Survey
Elmwood Avenue, Peltham, TW137AH, UK

ABSTRACT

The Digital Geographic Information Working Group was formed in 1983 and was established in response to the growing demands for defence geographic data. The DGIWG currently consists of 9 NATO nations but is not a NATO body and the standards it has created are equally applicable to the civil geographic community. A comprehensive set of exchange standards has been created, capable of eschanging raster, matrix and vector data (which may vary in intelligence from 'spaghetti' to topologically structured) on a variety of media. Particular care has been taken to ensure that supporting geographic reference and quality data can be passed with each dataset. A standardized form of coding the feature and attribute content of geographic data has also been developed. The family of DGIWG standards vary from being "Final Draft' to being published and under formal document change control.

INTRODUCTION

The Digital Geographic Information Working Group (DGIWG) was established in June 1983 with four nations as its members. The membership increased to six in 1984, seven in 1986, eight in 1989 and nine in 1990. The following countries are members: Belgium, Denmark, France, the Federal Republic of Germany, Italy, The Netherlands, Norway, the United Kingdom and the United States of America. In addition Canada and Spain are participating as observers and negotiating membership.

The formation of the DGIWG was originally in response to the growing demands from defence systems for digital geographic data. Many advanced aircraft, weapon, intelligence and information systems now need data of this type to enable them to carry out their roles; and many more systems with similar data requirements are under development. The introduction of defence systems of these types has clearly identified a requirement to standardise geographic data on an international basis to allow cooperative data capture and to ensure that the national production costs of digital geographic data are kept to a minimum. Although originally created to meet these defence needs, the DGIWG has been greatly influenced by civil standardization efforts for two reasons:

- Many national defence organizations rely heavily on the national civil mapping organizations for their data;
- In order to rationalize effort and capitalize on civil standardization work existing international standards are used wherever possible.

The comprehensive needs of defence users only mirror those of the civil community so that the DGIWG standards that have been derived are equally applicable to civil and defence environments. The DGIWG is keen to make its extensive standardization efforts available to the public domain in order to maximize the benefits of standardization and gain wide acceptance of the DGIWG standards.

SCOPE AND GENERAL GOALS OF THE STANDARDIZATION EFFORT

The prime goal of the DGIWG is to exchange digital geographic data, produced by DGIWG member nations, in accordance with agreed exchange formats, regulations, specifications and media. Time and effort in the Group is devoted mainly to aeronautical and land system geographic requirements, although hydrographic developments are kept under review by reports from the DGIWG committee members who work within their national hydrographic organisations.

Work within the Group is carried out under the terms of a formally agreed Memorandum or Understanding, an extract of which is given below:

DGIWG Memorandum of Understanding

Article 1: The aim of this agreement is to establish the general principles and regulations for the creation and maintenance of system allowing exchange of digital geographic information for defence use, among nations who produce such information of their territory or of other areas of NATO interest.

Article 2: Participating nations agree to:

1. Exchange, on request, digital geographic information they produce, in accordance with agreed exchange formats, regulations, specifications and media.

2. Provide qualified and authorised members to form a Digital Geographic Information Working Group, which will consist of:

   a. A Steering Committee.
   b. A Technical Committee.

3. Provide, where possible, qualified members for special sub-groups or other bodies as agreed by the Steering Committee, e.g. for a Panel of Experts.

Method of Working
The DGIWG works to an agreed development strategy. The basis of this strategy relies on the member nations committing themselves to the development and implementation of a multinational geographic information system. It has been assumed that this system will be formed of common standards and procedures, but without the need for individual nations to possess identical hardware or software. Furthermore, it has been accepted that the exchange of geographic information among nations should be possible without the use of a master database, and that the exchange of both data and specific geographic products will be agreed under bilateral arrangements within the procedures developed by the DGIWG.

In accordance with these general goals the DGIWG standards do not consider the architecture of national databases. The standards specify common approaches to the definitions of Features and Attributes, Values and Units and a Logical Structure of data elements and their relations. The exchange standards are designed to ensure that the receiver of data can integrate them into his own database and can understand the meaning of the digital geographic information.

BRIEF HISTORY AND BACKGROUND OF PAST EFFORT

As indicated in the Memorandum of Understanding, the DGIWG is composed of three main bodies: a Steering Committee, which is chaired by the United Kingdom; a Technical Committee, which is chaired by the United States of America; and Working Groups, which form a Panel of Experts who develop the standards and technical documentation required.

DGIWG meetings are held at locations which are rotated among the member nations. The Steering Committee meetings, which are normally held twice a year, are formal, with presentations of national positions leading to decisions on the development of exchange standards. The Technical Committee is tasked by the Steering Committee and provides recommendations as appropriate. The Panel of Experts report to the Technical committee which lays down its schedule and goals. The strength of this structure is in the work carried out by the Panel of Experts. Whereas the Technical Committee meets twice a year, the Panel of Experts meets as often as is necessary to achieve the schedule of work laid down by the

Technical Committee. Panel of Expert meeting scan last up to three or four weeks, and at times these Working Groups have met for total of ten to twelve weeks in particular one year periods. Thus a large volume of internationally agreed documentation has been generated over a relatively short time.

The work of the DGIWG over the last 6-7 years has concentrated on the standardization of the exchange of geographic data between mapping agencies. This has led to the standardization of the following aspects:

- o The data structures to be supported (including raster, matrix and vector (incorporating both chain-node and topologically structured data));
- o The data formats;
- o A Feature and Attribute coding scheme;
- o The exchange media (currently magnetic tape and CD-ROM);
- o The administrative procedures.

The types of data that have been considered for exchange include the digital representation of the following:

- o Elevation and depth information;
- o Geographic feature geometry and feature attribute information;
- o Information concerning the appearance and status of the earth's surface and its signature in the electromagnetic spectrum, eg radar, infra-red, etc.;
- o Ancilliary text information that may be linked to geographic features.

The data structures considered are capable of accommodating:

- o Neutral geographic data (captured without conforming to the requirements of a specific application);
- o Processed 'Product data' (potentially selected, generalized and transformed neutral data which may be combined with other user data);
- o Information about data holdings which may comprise:
  - Data Quality Information (Source, Accuracy, Consistency, Completeness, Currency, Security/Privacy Marking)

- Availability (Production Status, Releasability)

CURRENT WORK

Current work is concentrating on the testing of the exchange standards by the international exchange of digital data sets. At this time documentation has been produced to define a generic Digital Geographic Information Exchange Standard (DIGEST), a Feature and Attribute Coding Catalogue (FACC), two minimum raster data exchange specifications - Standard Raster Graphics (SRG) and ARC Standard Raster Graphics (ASRG), and two raster product specifications - Arc Standard Raster Product (ASRP) and Universal Transverse Mercator / Universal Polar Stereographic Standard Raster Product (USRP). Details of these documents are given below.

The Digital Geographic Information Exchange Standard (DIGEST)
DIGEST is a generic document, having been designed to establish a uniform method for the exchange of digital geographic data. It covers all types of exchange, including data in vector, raster and matrix formats. DIGEST'S uniformity is based on a common logical organisation of any geographic data set exchanged, whatever the data exchange structure used, and on a method of representation of the data which is currently taken from the International Standardisation Organisation (IS) 8211 Specification for Data Descriptive Files for Information Exchange. The following aspects are standardized:

- o Definition of the logical organization and structure of the data;
- o Definition of the information for interpreting the overall content of a digital geographic data transmittal;
- o Definition of the information supporting each specific geographic dataset;
- o A common method of identifying features and their descriptive attributes;
- o Statements of the quality and accuracy of the data;
- o Definition of the organization and representation of the data (ie format) on the exchange media;
- o Definition of the recording standard used for the exchange media.

Each of these aspects is discussed in more detail below.

Logical Organization: The following logical types of data subset are defined:

- o The Transmittal Header File (THF). This describes the

contents of the transmittal which may comprise one or several datasets encoded on one or several media volumes.

- o The Header Data Subset (HDS). This second data subset type describes the supporting information specific to each transmitted geographic dataset.

- o The Geographic Data Subset (GDS). The actual data of each transmitted geographic dataset.

The Transmittal Header File occurs once for each transmittal. The Header Data Subset and the Geographic Data Subset occur in pairs for each geographic dataset transmitted.

In accordance with ISO 8211 each of the above files contains one Data Descriptive Record and its companion Data Records. Each Data Descriptive Record contains the control parameters and data definition necessary to interpret companion data records. Each record is divided into fields and each field into subfields.

Transmittal Header File: This file occurs only once at the beginning of each transmittal and defines the total contents of the transfer. This may comprise one or several geographic datasets encoded on one or more media volumes. The file includes a definition of the Originator and Addressee, together with the data structure (raster, vector etc.), specification and overall geographic coverage of each geographic dataset transmitted.

Header Data Subset. This file describes the supporting information for each Geographic Data Subset and may contain the following files:

- o General Information File. This includes dataset identification, parameters (eg. structure, resolution, feature counts) and tiling system.

- o Geographic Reference File. This includes projection parameters, registration points and diagnostic points.

- o Source File. This includes information about the source documents relevant to the Geographic Data Subset, including Geodetic Datu, Vertical Datum, Source Projection, Inset identification and Legend data.

- o Quality File. This includes information about currency, consistency and completeness, attribute accuracy, positional accuracy and free text comments.

Geographic Data Subset. The actual geographic data transmitted may be vector (topological, chain-link or spaghetti), matrix (value or topological data is illustrated in Figure 1. The Exchange Structure Scheme for chain-link

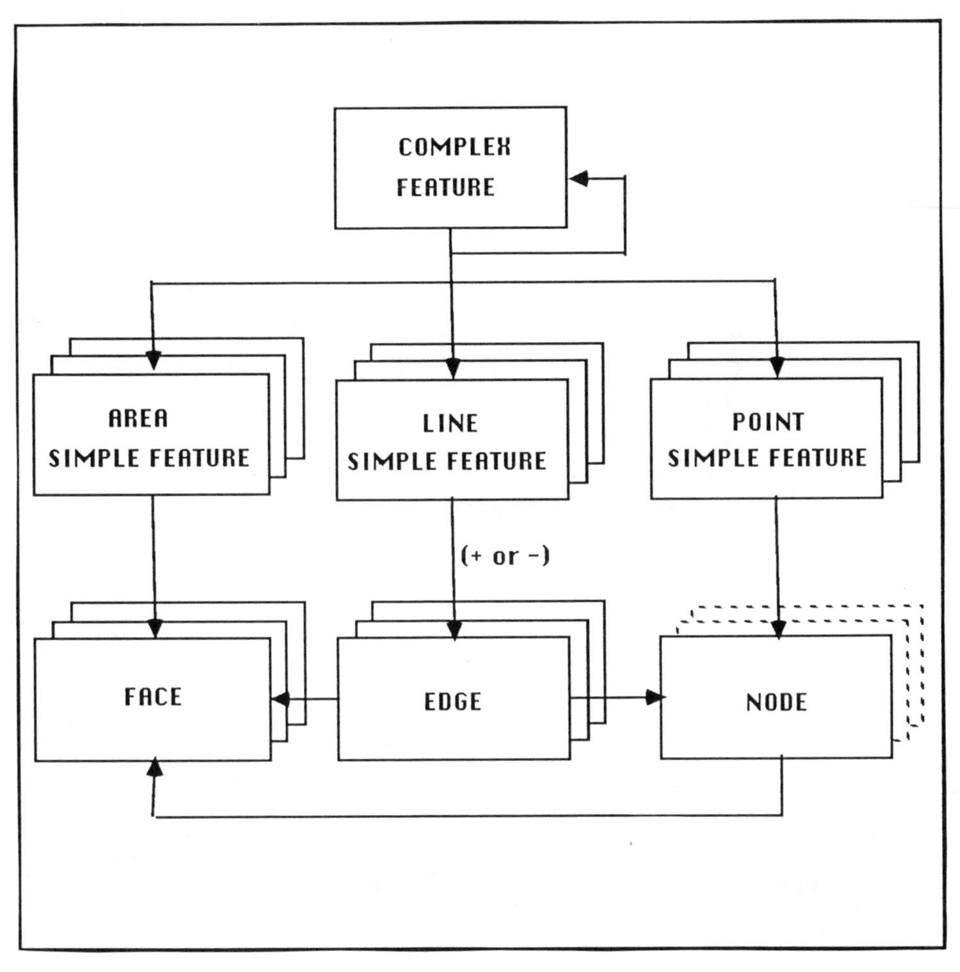

Figure 1. Exchange structure scheme for topological vector data.

and spaghetti data is illustrated in Figure 2. The basic logical structure of matrix and raster data is identical and is illustrated in Figure 3.

Feature and Attribute Coding. The Feature and Attribute Coding Catalogue (FACC) is a major appendix to DIGEST. It provides a common menu of geographic features and attributes, including a wide range of hydrographic information, along with a standardized coding system, to meet the comprehensive needs of geographic information systems. A global scope was considered in the definition of features and attributes. FACC is not a product, nor was it developed to the requirement of any single geographic application. Thus in order to define specific digital products FACC must be employed in concert with product specifications. These

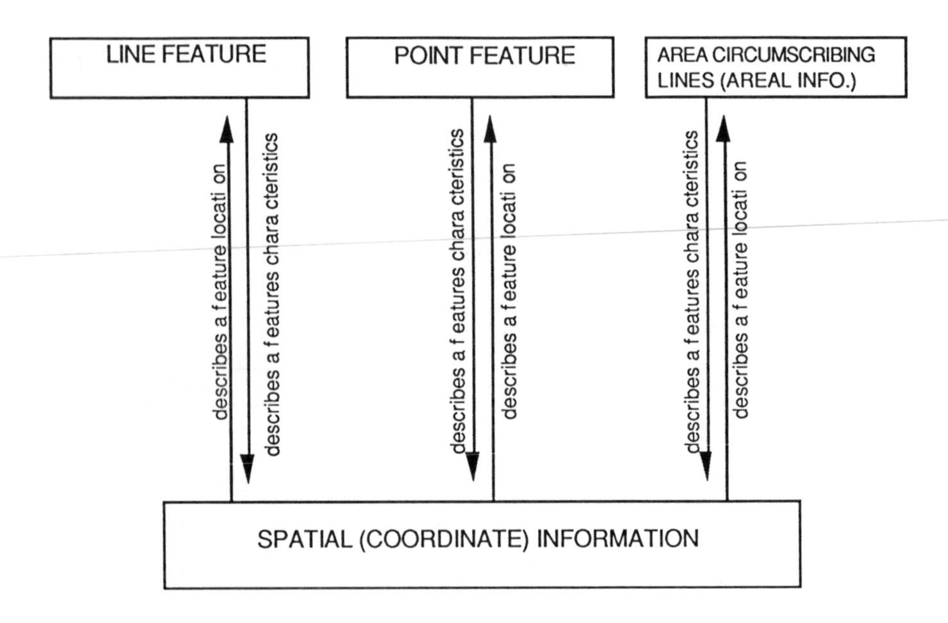

Figure 2. Exchange structure scheme for chain link and spaghetti data.

product specifications determine which features and attributes will be collected as well as defining specific collection criteria, such as positional accuracy and feature granularity. As FACC is tied to no specific data base of product line, it can be easily modified and updated in an onwards compatible manner to keep pace with dynamic technology improvements and the refinement of data requirements.

FACC is a vital standard since, without a common understanding of data content between systems, all considerations of structure and format become meaningless. The FACC is available as a fully published standard and is subject to formal Document Change control Procedures.

Data Quality. Quality information relevant to the overall Geographic Data Subset has already been described in the Header Data Subset. Data Quality Descriptors may be assigned in the geographic data when very detailed quality information is required at the Feature and Attribute level.

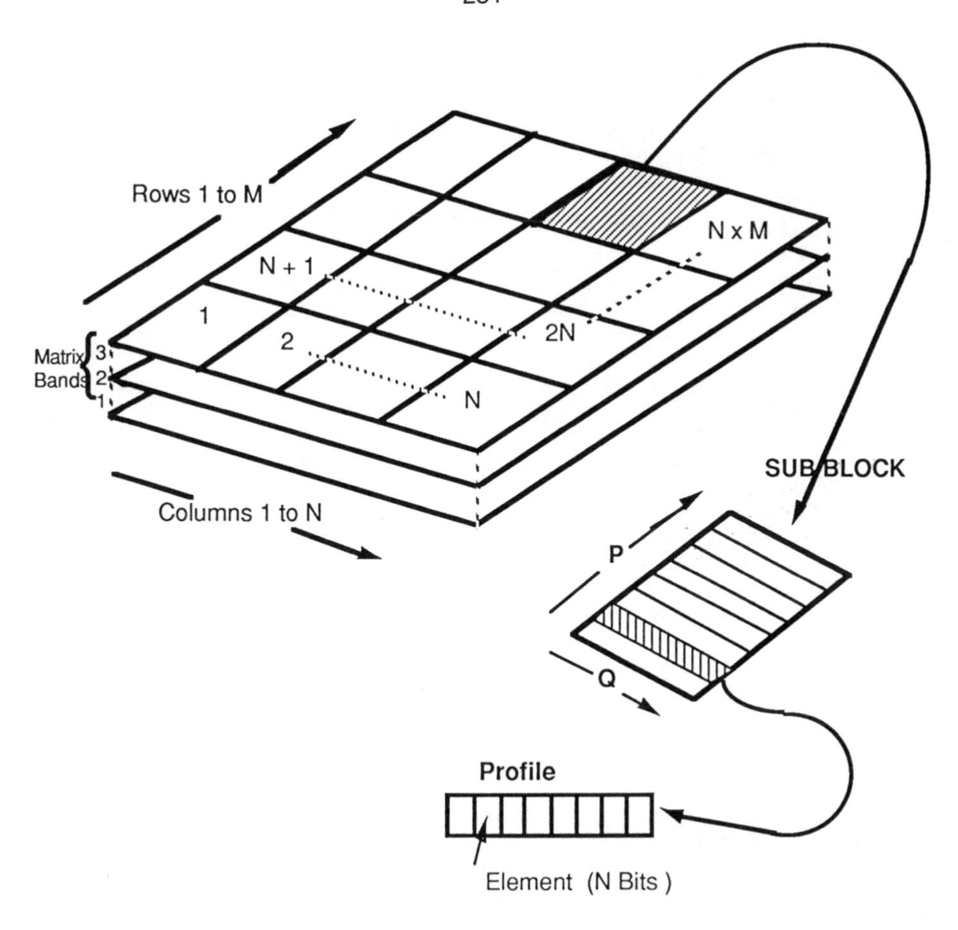

Figure 3. Matrix data set logical structure.

Exchange Media. The method of character set representation is standardized by adopting ANSI X3.4-1977 (ISO 646 IRU) with extensions conforming to ISO 2022. International standards are currently defined for Magnetic Tape (6250 GCR and 1600 PE) and CD-ROM transfer. It is intended to address other media and electronic communications requirements as the need arises.

Other Standardization Work

The DIGEST has been designed to support generic geographic data exchange. The national mapping agencies within DGIWG need to exchange specific data to support agreed joint production and exchange programmes. This has led to a family of standards based on DIGEST but limiting many of the generic parameters to defined ranges. These minimum exchange standards and product standards are described below.

Standard Raster Graphics (SRG). SRG is a minimum raster exchange standard for raster data allowing the exchange of digital facsimilies of graphic products. Red, green and blue (RGB) components, or colour-coded representation may be used. The total format of a graphic including margin, border and legend areas is scanned at a minimum resolution of 100 microns dependent upon the digitising device and the graphic to be scanned. considerable development work has been put into the definition of this resolution which ensures the high quality display of cartographic raster data. The resultant data consists of one graphic product which is provided with geographic control on the original datum and projection so that transformation and seaming of the data can be performed at the user's discretion.

ARC Standard Raster Graphics (ASRG). ASRG is a minimum raster exchange standard enabling the exchange of data in a seamless manner. The Equal Arc-Second Raster Chart (ARC) system is the projection and coordinate system for all ASRG data. The design objective of ARC is to provide graphic data in a virtually seamless manner and permit direct display in a nearly conformal presentation. The ASRGs consist not only of transformed graphic data but also a record that contains datum shift and projection parameters with which users can transform ASRGs back to the source graphic's datum or projection. ASRG data may consist of RGB or colour-coded images with a nominal 100 microns sample size, trimmed at the graphic's neat line and transformed to the ARC system frame of reference. Data on the ARC system will be maintained as a world-wide seamless data base of scanned graphic data on World Geodetic System 1984 (WGS 84).

Arc Standard Raster Product. This product standard defines a raster dataset which conforms to ASRG but in which the generic parameters have been fixed to allow the specification of a global seamless raster electronic display product. To reduce data volumes and to facilitate manipulation of the image on display a colour coded pixel representation has been adopted.

Universal Transverse Mercator/Universal Polar Stereographic Standard Raster Product (USRP). This product standard has been designed to support global display applications on the UTM/UPS projections as opposed to the Arc Projection. With the exception of projection the product is specified in an identical fashion to ASRP.

PLANNED WORK

In addition to the testing programme currently underway, the DGIWG is starting to consider common international vector product and neutral data requirements and is planning the generation of specifications to standardize these datasets. Significant advantages can be gained in the exchange of products since these remove the post processing requirements for specific applications which become necessary with the exchange of neutral data. It is hoped that future work in this area may lead to other common product specifications being developed and agreed on a wide international basis. The emphasis of DGIWG work is therefore shifting to place a greater emphasis on user oriented as opposed to producer oriented standards.

The DGIWG is also beginning an examination of telecommunication requirements for digital geographic information. It is possible that the exchange of data over communication networks and the use of data disks as interactive sources of data means that the current exchange standards, based on ISO 8211, may be inadequate for these purposes. A number of other international standards, notably ISO 8824/5, may need to be considered, and the DGIWG documentation expanded to take account of the telecommunication aspects of these. Work over the next two years, lead by Canada, should resolve this matter.

The evolution of the DGIWG standards is being planned in a controlled manner to ensure that emerging Geographic Information Systems requirements can be met using the same overall logical structure and to maximize onwards compatibility as the standards are expanded.

Finally, DGIWG is extending its contacts with other geographic agencies. Links with the International Hydrographic Organisation and the International cartographic Association are being explored with a view to sharing ideas on areas of technical interest, perhaps leading to mutual cooperation to ensure compatibility of approach. this will reduce the production costs of individual mapping agencies and reduce the proliferation of exchange standards that users are currently facing.

AVAILABILITY OF REPORTS AND STANDARDS

DGIWG documentation is available as follows:

- o DIGEST available as a Final Draft while live data testing continues);
- o FACC (approved as a published standard and has been placed under Document Change Control);
- o SRG, and ASRG (available as Final Drafts);
- o ASRP (available as a pre-Final Draft).

The testing programme which is in progress may lead to minor amendments to the documentation and rationalization of the raster documentation (SRG, ASRG, ASRP and USRP) is being considered. Copies of both the draft and published standards can be obtained from The Secretary DGIWG, Directorate General of Military Survey, Elmwood Avenue, Feltham, Middlesex, TW13 7AE, UK.

SELECTED BIBLIOGRAPHY

The DGIWG standards reference the following other standards:

American National Standards Institute (ANSI), 1973, ANSI X3.39-1973, "Recorded Magnetic Tape for Information Interchange" (1600 CPI,PE).

American National Standards Institute (ANSI), 1976, ANSI X3.54-1976, "Recorded magnetic Tape for Information Interchange" (6250 CPI, Group-Coded Recording).

American National Standards Institute (ANSI), 1977, ANSI X3.4-1977, "Code for Information Interchange".

American National Standards Institute (ANSI), 1978, ANSI X3.27, "Magnetic Tape Labels and File Structure for Information Interchange (ASCII)".

Federal Information Processing Standard (FIPS), 1980, FIPS PUB 1-1 "Code for Information Interchange".

Federal Information Processing Standard (FIPS), 1973, FIPS PUB 25, "Recorded Magnetic Tape for Information Interchange," (1600 CPI, Phase Encoded).

Federal Information Processing Standard (FIPS), 1978, FIPS PUB 50, "Recorded Magnetic Tape for Information Interchange", 6250 CPI (246CPMM) Group Coded Recording.

Federal Information Processing Standard (FIPS), 1980, FIPS PUB 79, "Magnetic Tape Labels and File Structure for Information Interchange".

International Standards Organization (ISO), 1985, ISO 8211, "Information Processing: Specification for a Data Descriptive File for Information Interchange".

International Standards Organization (ISO), 1988, ISO 9660, "Information Processing: Volume and File Structure of CD-ROM for Information Interchange".

The following work has also been used:

Moellering, H. (Ed), 1986, Issues in Digital Cartographic Data Standards, Report No 7, National Committee for Digital Cartographic Data Standards.

Moellering, H. (Ed), 1988, Issues in Digital Cartographic Data Standards, Report No 8, National Committee for Digital Cartographic Data Standards.

Acknowledgement

Much of the content of this chapter was contributed by Lt. Col. Peter Walker, Secretary of the DGIWG 1986-90.

INDEX OF CONTRIBUTORS

SUBJECT INDEX